Soft Matter

Ashim Kumar Dutta

Soft Matter

Fundamentals and Applications

Springer

Ashim Kumar Dutta
Research and Development
India Glycols Limited
Kashipur, Uttarakhand, India

ISBN 978-981-96-0623-8 ISBN 978-981-96-0624-5 (eBook)
https://doi.org/10.1007/978-981-96-0624-5

This Springer imprint is published by the registered company Springer Nature Singapore Pte Ltd.
The registered company address is: 152 Beach Road, #21-01/04 Gateway East, Singapore 189721, Singapore

I would like to dedicate this book to the loving memory of my mom and dad to whom I owe everything I am today.

Preface

Over the past three decades, there has been considerable interest on materials that we refer to as "condensed soft matter" or "complex fluids". As the name suggests, these materials are readily deformed under small perturbative forces and have properties intermediate to that of hard solids and liquids. Polymers, Liquid Crystals, Colloids, Gels, Foams, Emulsions, Slurries and many more belong to this family of materials. Because several of our daily use essentials like food, cosmetics, personal care items belong to this class of materials, this branch of study is therefore of interest to scientists in academic and industry. Although a large volume of literature by experts already exists, we still felt a need of a simple and concise book that would assist researchers new to this subject get a broad overview and help get started on a journey into research and innovations in this fascinating field. It is again for the same above-mentioned reasons, the language has been kept simple and straightforward and complicated mathematical deductions has been avoided wherever possible. However, appropriate references have been provided for those who would be interested into the derivation of these mathematical equations. I anticipate that this book will be of interest to researchers in chemistry, physics, biochemistry, foods, materials chemistry, pharmaceuticals and chemical engineers. Chapter 1 introduces to the reader the fundamental aspects of soft matter and its difference with conventional solid state condensed matter. A brief overview of the various intermolecular forces contributing to self-assembly of the molecules in these systems is briefly discussed. Chapter 2 is a recapitulation of classical thermodynamics, a cornerstone of our understanding of the fundamental principles of science. The four state functions Internal energy (U), Enthalpy (H), Helmholtz Free Energy (F) and Gibbs Free Energy (G) and their applications in phase transformation, homogeneous and heterogeneous nucleation and crystal growth is covered in details. Chapter 3 focusses on the physics of various interfacial phenomena like self-assembly of molecules at interfaces that results in lowering of surface tension, adsorption, adhesion and spreading of liquids on surfaces that form the heart of product design across diverse industries ranging from detergents to pharmaceuticals. Chapter 4 is a compilation of various experimental techniques employed in the examination of physicochemical properties of various soft condensed matter systems. Interfacial properties by tensiometry and imaging while bulk properties by thermal and rheological measurements has been discussed in some details. Chapter 5 provides an introduction to surfactants and their self-assembling properties in bulk and at the air-water interface. Temperature and concentration dependent properties of various surfactants is briefly discussed. Chapter 6 presents an overview on emulsions, their physico-chemical properties and conditions required for their stability. The role of surfactants, polymers and particles in stabilizing emulsions has been covered in details. Chapter 7 entitled hydrogels is an emerging area that find many applications across industries like foods, cosmetics and advanced technological areas like regenerative medicine, tissue engineering and organ growth and its replacement. In this chapter, the fundamental aspects of hydrogels have been discussed with a focus on materials that form hydrogels. Chapter 8 presents various applications across industries and new developments to provide a flavor of this fascinating emerging area especially those new to this area of research and innovation and also be a ready reference for an expert.

The author takes this privilege to thank Prof. T. N. Misra of Indian Association of Cultivation of Science, Kolkata, India for introducing me to the fascinating world of Soft Matter and its many applications. Last not least, I will like to thank Malabika, my wife and kids Sourav and Siddharth for their cooperation and bearing with me while I worked on weekends.

Kashipur, India Ashim Kumar Dutta

Contents

Chapter 1
What Is Soft Matter?

1.1 Introduction

Matter consists of atoms and molecules that are packed differently to form three well-known states of matter -solids, liquids, and gases that we are familiar with from our school days. Atoms and molecules have a mean diameter of about 10^{-9}–10^{-10} m while the human eye has a resolution of 1×10^{-6} m and hence are invisible to the human eye. Although an individual atom or molecule is invisible, objects in the real world around us that we can see and touch consist of trillions of these invisible entities. It is the difference in arrangement, packing, and strength of interaction between molecules that makes a material hard like steel or soft like flower petals. For decades, the structure and properties of solids namely metals, crystals, glasses, and semiconductors have been studied as main topics in solid state and condensed matter physics and although soft materials like food, biopolymers, and biological membranes are known for centuries, the terminology 'Soft Matter' had never been used. In 1991, in his Nobel Prize Lecture, Prof. Pierre Gilles De Gennes used the term "Soft Matter" for the first time to describe materials that undergo large deformations under small perturbative forces [1]. Ever since, this terminology has been used to describe materials like soaps, gels, polymers, liquid crystals, emulsions, colloids, tissues, and muscles among others. An alternative terminology "Complex Fluids" has also been used to describe these materials that stand justified because the properties of these materials are intermediate to that of hard solids and liquids. While metals and crystals clearly belong to the family of hard crystalline solids, others namely colloids, liquid crystals, polymers, gels, waxes, emulsions, tissues, and muscle fit into the descriptions of soft condensed matter [2, 3].

Today, soft condensed matter has emerged as an area of active research and innovations that makes it interesting to physicists, chemists, biologists, materials chemists, and engineers alike as it covers a wide range of totally diverse subjects like food, biological cellular matter, gels, liquid crystals, emulsions and polymers that are of academic and industrial interest. In this chapter, we briefly discuss various aspects

A. K. Dutta, *Soft Matter*, https://doi.org/10.1007/978-981-96-0624-5_1

of soft matter and its inherent difference from conventional condensed solid-state materials.

1.2 Building Blocks of Soft Matter

Single atoms and molecules are undoubtedly the smallest entities that makeup matter. In the gas phase, these atoms and molecules are far apart from each other and interaction between them is negligible, hence each atom or molecule acts like a free particle. In liquids, the distance between the molecules is closer, and in solids, the molecules are tightly packed which is manifested through large differences in their density, melting and boiling points, thermal expansion coefficients, and modulus of elasticity among others. Two approaches have been adopted by various workers to study materials and correlate their macroscopic properties to their molecular composition and organization. While one conventional approach is “Top Down”, the other is “Bottom Up” as shown in Fig. 1.1. Although in principle both approaches are equivalent, the “Bottom Up” approach is considered more insightful as it allows building the 3D macroscopic structures molecule by molecule provides a deeper understanding of the changing macroscopic behavior of materials with increasing population of molecules in the system and transitions from 1 to 2D and 3D.

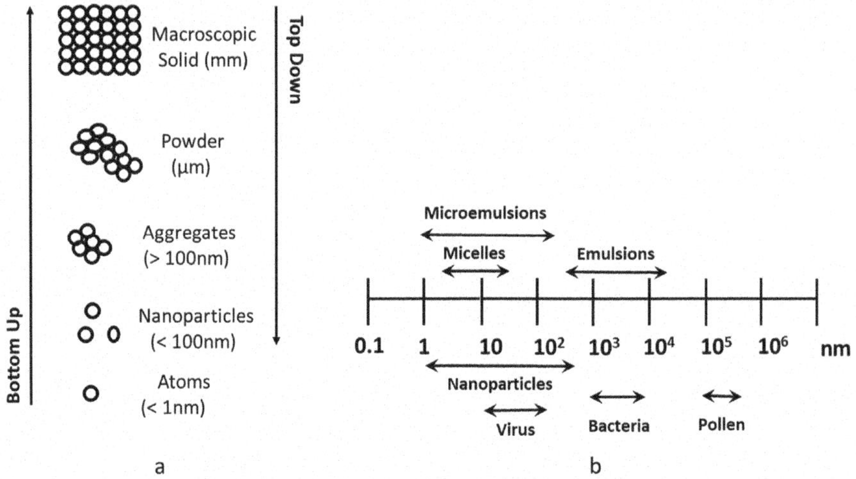

Fig. 1.1 **a** Schematic of “top down” and “bottom up” approaches to problem-solving. **b** Length scales of various soft matter systems

1.3 Characteristic Features of Soft Matter

Reference to a material being "hard" materials brings to mind materials like metals, minerals, alloys, and many others. In contrast, "soft" materials exhibit a large response to small perturbative forces. Students of physics are accustomed to the terminology "condensed matter" where solid-state physics, physical properties of metals, semiconductors, and alloys are taught [3, 4]. Such materials have some unique features, namely a well-defined crystalline lattice structure, anisotropic optical and elastic properties, and phase transition temperatures. Soft materials on the other hand have properties that are distinctly different as discussed below.

1.3.1 Elasticity

The modulus of elasticity (ε) defined as the ratio of stress (σ) to strain (γ) is high for hard solids and significantly low by several orders of magnitude for soft matter materials [2, 5]. A worthwhile example perhaps is a comparison of the Young Modulus (E) and percentage elongation of mild steel and isoperene rubber. While for mild steel and isoperene rubber, the Young Modulus (E) is 210 and 35 GPa, the percentage elongation for mild steel and isoperene rubber is 35% and 510%, respectively.

1.3.2 Viscoelasticity

Hard solids are rigid and do not flow and hence do not have viscosity. On the contrary, soft solids have properties intermediate to that of solids and liquids and hence exhibit viscoelastic properties which are often nonlinear [2, 4, 5]. Measure of storage modulus (G') and loss modulus are measure of its elastic (solid $\frac{G'}{G''} > 1$ represents a gel while $\frac{G'}{G''} < 1$ represents a flowable liquid [6–8]

1.3.3 Mesoscopic States

In simple terms, a system is said to be in equilibrium if its state and properties do not change over long periods of time especially during the time frame of its evaluation [8]. Thermodynamically, such a state is achieved when the system is at the lowest point in the potential energy curve (Fig. 1.2). Hard solids like metals, alloys, and ceramics do not exhibit change in their properties over very long periods of time, often years if stored appropriately, and hence are considered to have attained a state of mechanical, thermal, and thermodynamic equilibrium. Soft matter materials like liquid crystals, gels, and many others show changes in their physico-chemical

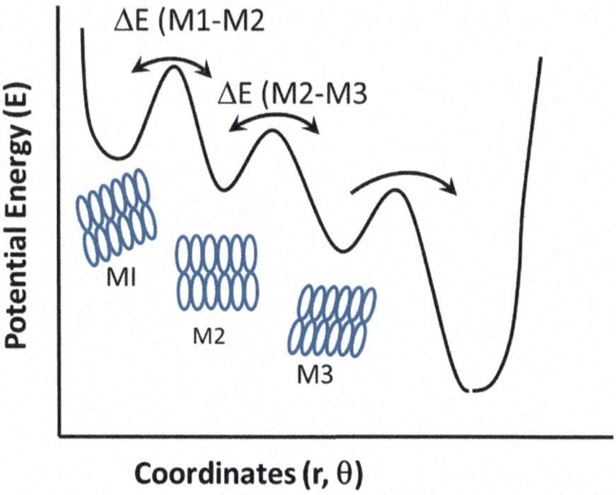

Fig. 1.2 Schematic of potential energy versus coordinates. $M1$, $M2$, and $M3$ are the mesophases. Difference in energy between $\Delta E(M1 - M2)$ and $\Delta E(M2 - M3)$ is k_BT

properties at ambient over time that may range from a few hours to months [8]. These observations have been attributed to the weak intermolecular forces between the molecules that allow for displacement of molecules over small distances and rotation in 3D space that is reflected in changes in its optical and viscoelastic properties. These configurational changes correspond to different potential energies that give rise to different mesoscopic states and the energy barrier between these different mesoscopic states are low of the order of thermal energy $k_BT \sim 4.2 \times 10^{-21}$ J at 300 K. This feature permits transitions from one mesoscopic state to another that occurs over very long periods of time [8, 9]. Transitions from mesoscopic states $M1$ to $M2$, $M2$ to $M3$, and vice-versa becomes possible which explains changes in its physico-chemical properties when left to itself. Because of the low-energy barrier between various mesoscopic states, changes are readily observed due to small changes in temperature or externally applied fields (electric and magnetic) mechanical stresses, and thermal changes. Such transitions result in very different morphologies and macroscopic properties of the material achieved through different kinetic pathways.

1.3.4 Order–Disorder

In crystalline solids, a long-range order exists where the unit cell structure is repeated over length scales that are several orders greater than the unit structure. X-ray diffraction patterns of such ordered materials show sharp peaks that may be used to determine its lattice structures. In crystalline solids, the lattice energy is large which is manifested as large latent heat of melting and high melting point. On the contrary, in soft materials thermal energy (k_BT) is sufficient to perturb the system, and because the energy required is minuscule changes occur spontaneously when the materials are left in waiting due to fluctuations in the temperature of the ambient [2, 3, 5, 8, 9].

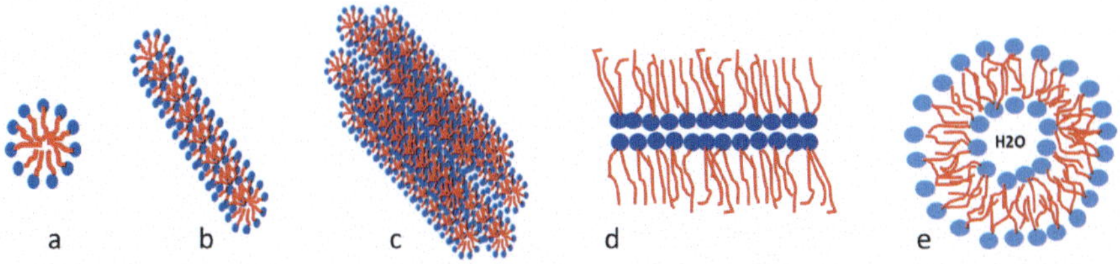

Fig. 1.3 Schematic of **a** spherical micelle, **b** rod micelle, **c** hexagonal phase (*Hα*), **d** bilayer and **e** vesicle. Blue-filled circles are hydrophilic groups and Red chains are hydrophilic long chains

1.3.5 Self-assembly

This is a spontaneous process where atoms or molecules arrange themselves to form specific structures that are thermodynamically permissible. Excellent examples are the formation of spherical micelles (Fig. 1.3) by surfactants above its critical micellar concentration (CMC), as discussed in detail in chapter five of this book. With increasing concentration above CMC, different structured entities like cylindrical micelles or complex cubic phases are developed that are manifested through large changes in viscosity, refractive index, solubility, and birefringence among other properties [5–8]. Similarly, naturally occurring phospholipids consisting of a hydrophilic head group and hydrophobic double chains when dispersed in water form bilayers (Fig. 1.3) that are essential parts of biological membranes, vesicles that play important physiological roles in living systems [7–12].

1.4 Molecular Interactions

Conceptually starting with a single atom or molecule and adding one at a time results in the formation of dimers, trimers, and small clusters are formed that grow with increasing number of atoms or molecules added to form large structures that may be liquids, amorphous or crystalline solids depending on the intermolecular forces that bind them together. A large fraction of these forces are of electrical origin as briefly discussed in the section below [13–16].

1.4.1 Ion–Ion

Like charges repel and unlike charges attract is something we all know from our high school days and the net force of attraction or repulsion is governed by Coulomb's law that is mathematically expressed as

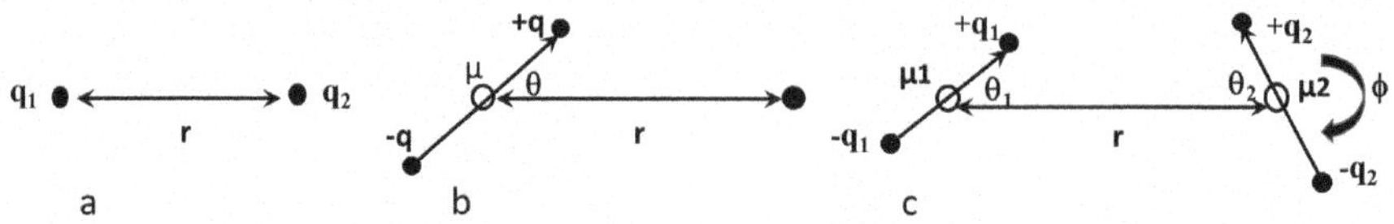

Fig. 1.4 **a** Interaction between two point charges, **b** interaction between a point charge and a dipole, **c** interaction between two dipoles

$$F_C = \frac{q_1 q_2}{4\pi \varepsilon_0 \varepsilon_r r^2} \tag{1.4.1.1}$$

where ε_0 and ε_r are dielectric constant of vacuum and the relative dielectric constant of that medium in which the charges q_1 and q_2 are located (see Fig. 1.4a).

$$\text{Since } F = -\frac{\mathrm{d}U(r)}{\mathrm{d}r} \text{ we get } U_C = \frac{q_1 q_2}{4\pi \varepsilon_r \varepsilon_0 r} \tag{1.4.1.2}$$

where $U_C(r)$ is the electric potential. It must be noted that when charges q_1 and q_2 are similar then their product is always positive suggesting repulsion and when the charges are dissimilar then their product is negative suggesting that the force is attractive. Typically, the energies associated with such interactions between ion pairs is about 100×10^{-20} J which is about 250 times higher than the thermal energy $k_B T$ (4.2×10^{-21} J).

1.4.2 Ion–Dipole

Although atoms and molecules have net zero charge, nevertheless departure of the electron cloud from symmetry about the nuclear charge causes a pair of small charges ($+q$ and $-q$) separated by distance (l) is created. Such a configuration corresponds to a dipole having a dipole moment (μ) where $\vec{\mu} = q\vec{l}.\vec{\mu} = q\vec{l}$. Interaction potential between a single ion with charge (q) and a fixed dipole separated by distance r as shown in Fig. 1.4b is given by 1.2b1.

Considering a freely rotating dipole, the equation modifies to

$$U_{\text{ion-dipole}} = -\frac{q^2 \mu^2}{6(4\pi \varepsilon_0 \varepsilon_r)^2 r^4 k_B T}. \tag{1.4.2.1}$$

The r^{-4} dependence shows that the interaction strength decreases more rapidly and the inverse dependence on temperature indicates that the interaction weakens with increasing temperature as well.

1.4.3 Dipole–Dipole

When two dipoles are located such that their centers are at a distance r and the dipole moments μ_1 and μ_2 make angles θ_1 and θ_2 with the line joining the centers of the dipoles (Fig. 1.4c) and ϕ is the angle between the projections of the dipole to an area perpendicular to the connecting line. The dipole–dipole potential energy is given as

$$U_{D-D} = -\frac{\mu_1\mu_2}{4\pi\varepsilon_r\varepsilon_0 r^3}[2\cos\theta_1\cos\theta_2 - \sin\theta_1\sin\theta_2\cos\phi] \tag{1.4.3.1}$$

and considering the dipoles freely rotating in space we have the Keesom potential

$$U_{D-D} = -\frac{\mu_1^2\mu_2^2}{3(4\pi\varepsilon_r\varepsilon_0)^2 r^6 k_B T}. \tag{1.4.3.2}$$

The r^{-6} dependence shows that the interaction strength decreases more rapidly and the inverse dependence on temperature indicates that the interaction weakens with increasing temperature as well.

1.4.4 Ion-Induced Dipole

When a charged particle q_1 approaches a neutral molecule without a static dipole moment, no interaction is expected. However, if the electric field created by the charge causes polarization of the molecule generating a dipole moment μ then the potential energy of the interaction is

$$U_{\text{ion-Induced-Dipole}} = -\frac{q_1^2\alpha_2}{2(4\pi\varepsilon_r\varepsilon_0)^2 r^4}. \tag{1.4.4.1}$$

Here α is the polarizability of the molecule and $\mu = \alpha E$ where E is the electric field generated by the charge q_1. Considering freely rotating dipoles we have the Debye potential

$$U_{\text{ion-Induced-Dipole}} = -\frac{\mu_1^2\alpha}{(4\pi\varepsilon_r\varepsilon_0)^2 r^6}. \tag{1.4.4.2}$$

1.4.5 London Dispersion Interactions

All the calculations above are based on classical physics and interaction between charges are based on Coulomb's law of electrostatics. These however does not explain the interaction between non-polar molecules. This riddle was however solved by Fritz London who argued that electrons moving about the nucleus in an orbit generate a circuital current that is equivalent to an electric dipole. The fact that the electron moves with a high frequency (10^{15}–10^{16} Hz) will cause the dipole moment to change with the same frequency. Considering a quantum mechanical approach London derived an expression

$$U_{\text{London}} = -\frac{3}{2}\frac{h\nu_1\nu_2\alpha_1\alpha_2}{(4\pi\varepsilon_r\varepsilon_0)^2(\nu_1+\nu_2)r^6} \tag{1.4.5.1}$$

that shows a r^{-6} dependence like the Debye and Keesom potentials.

1.4.6 Van Der Waals Forces

These are the sum total of inductive and dispersive long-range intermolecular forces that contribute significantly toward the stability of the systems despite being weak forces typically **0.5–15** kJ/mole that arise from dipole–dipole interactions between molecules. There are three components that contribute to the van der Waals forces (a) permanent dipole–dipole interaction, (b) permanent dipole–induced dipole interaction, and (c) induced dipole–dipole interaction.

Permanent Dipole–Dipole Interaction: Considering a distribution of randomly oriented molecules with dipole moments (μ_i) Keesom derived an expression for the potential energy as

$$U_{\text{Keesom}}(r) = -\frac{\mu_i^4}{k_B T r^6}. \tag{1.4.6.1}$$

Clearly, the interaction is attractive and directly proportional to the fourth power of the dipole moment and inversely proportional to r^{-6}.

Dipole-Induced Dipole Interaction: Permanent Similarly, Debye considered the interaction between a molecule with a permanent dipole moment (μ_i) inducing a dipole moment ($\alpha_i\mu_i$) in another molecule derived the expression of potential energy as

$$U_{\text{Debye}}(r) = -\frac{\alpha_i\mu_i^2}{r^6}. \tag{1.4.6.2}$$

Because the electron cloud in an atom or molecule is not stationary but fluctuates that causes perturbation in the electron cloud in the neighboring atom or molecule. London considering the problem from a purely quantum mechanical perspective provided an expression for the potential energy as

$$U_{\text{London}}(r) = -\frac{3}{4}\frac{\alpha_i^2 h\nu}{r^6}. \tag{1.4.6.3}$$

Thus the net van der Waals potential is a sum total of the above three components. It must be noted that all the potentials are attractive in nature as indicated by the negative sign and all of these has a dependence on r^{-6}.

1.4.7 Hydrogen Bonding

These are weak attractive forces the magnitude of which lies intermediate to covalent and Van der Waals forces and is typically of the order **2–6 x 10^{-20}** J which is about 5–15 $K_B T$, It accounts for many of the peculiar and anomalous properties of water and plays a crucial role in self-assembly molecules in bio-membranes and other supramolecular structures. The double helix structure of DNA and its structural stability have been attributed to base pair hydrogen bonding. Although it is 20 times weaker than covalent forces, it is 50 times stronger than van der Waals interactions making the contribution from H-bonding significant [15]. Have successfully demonstrated that H-bonding interaction potential may be expressed as

$$U_{HB} = \frac{A}{r^6} + \frac{B}{r^8} + \frac{C}{r^{10}} + \frac{D}{r^{12}} + \frac{Eq_1 q_2}{\varepsilon r} \tag{1.4.7.1}$$

where A, B, C, D, and E are constants.

1.4.8 Hydrophobic Forces

A hydrophobic molecule when placed in water causes the water molecules to redistribute around it causing the free energy of the system to increase. Since thermodynamically, a stable system should have minimal free energy, aggregation of the hydrophobes occurs such that contact with structured water molecules surrounding the hydrophobe is minimized. In fact, the number density of water molecules near the hydrophobic substrate is lower. It must be also noted that the strength of hydrophobic interactions increases with temperature. Molecules with a larger number of hydrocarbons show stronger hydrophobic interaction. Surprisingly, aliphatic hydrocarbons show stronger hydrophobic interactions compared to aromatic hydrocarbons. The hydrophobic interaction force is attractive and is expressed as

$$F_{\text{Hydrophobic}} = -C\exp\left(-\frac{r}{\lambda}\right) \tag{1.4.8.1}$$

where C is a constant, r is the distance, and λ is the length over which the interaction dominates [15, 16].

References

1. de Gennes PG (2003) Simple views on condensed matter. World Scientific, N.J.
2. Jones RAL (2002) Soft condensed matter. Oxford University Press, N.Y.
3. Chaikin P, Lubensky T (2006) Principles of condensed matter physics. Cambridge University Press, U.K.
4. Warlimont H, Martienssen W (eds) (2018) Springer handbook of materials data. Springer, Germany
5. Terentjev EM, Weitz DA (2015) The oxford handbook of soft condensed matter. Oxford University Press, U.K.
6. Doi M (2013) Soft matter physics. Oxford University Press, U.K.
7. Hamley IW (2007) Introduction to soft matter: synthetic and biological self-assembling materials. John Wiley, U.K.
8. Hirst LS (2020) Fundamentals of soft matter science. Taylor & Francis, Boca Raton, USA
9. Nagel SR (2017) Rev Mod Phys 89:025002
10. Lindoy LF, Atkinson IM (2000) Self-assembly in supramolecular systems. Royal Society of Chemistry, Cambridge
11. Bohidar HB, Rawat K (2017) Design of nanostructures: self-assembly of nanomaterials. Wiley-VCH, Germany
12. Birdi KS (1999) Self-assembly monolayer structures of lipids and macromolecules at interfaces. Kluwer Academic Publishers, N.Y.
13. Ninham BW, Lo Nostro P (2010) Molecular forces and self assembly: in colloid, nano sciences and biology. Cambridge University Press, U.K.
14. Israelachvili J (1992) Intermolecular and surface forces. Academic Press, N.Y.
15. Zhou L (2019) Introduction to soft matter physics. World Scientific, Tokyo
16. Butt H-J, Kappl M (2010) Surface and interfacial forces. Wiley-VCH, Germany

Chapter 2
Classical Thermodynamics—A Primer

2.1 Introduction

Historically, classical thermodynamics has evolved over centuries that may be traced back to 1650 A.D. when Otto von Guereck designed the first vacuum pump in the form of Magdeburg spheres [1]. The term "Thermodynamics" is believed to originate from two Greek words "Thermis" meaning heat and "Dynamis" meaning power which is the science of heat energy and its transport. In 1656 A.D., Sir Robert Boyle and Sir Robert Hooke's pioneering work established the equation of state of gases as we know them today. In 1679 A.D., Denis Papin designed a high-pressure digester, a predecessor to the steam engine that was subsequently designed by Sir Thomas Savery and improved upon by Sir Thomas Newcombe in 1712 to become the first steam engine. A century later, in 1824 Sir Sadi Carnot in his famous book "Reflections on the Motive of Fire" provided a detailed design of the Carnot engine and the principles of the Carnot cycle that forms the conceptual cornerstone of modern thermodynamics. Between 1850 and 1876 Sir William Rankine, Sir William Thomson (Lord Kelvin), Sir Rudolph Clausius, and Sir Joshua Willard Gibbs formulated the three laws of thermodynamics, the concept of entropy and Gibb's equation which form the pillars of classical thermodynamics that finds diverse applications in physics, chemistry, biology and engineering [1, 2].

Classical thermodynamics is about energy transformation in macroscopic systems at equilibrium that has evolved from observations and careful experimentation and has stood the test of time. To date, violation of any of the laws of thermodynamics is not known. However, none of the above principles can be derived from any other principle. Despite the great successes of classical thermodynamics as an emerging and reliable science, the missing part of the jigsaw puzzle was the connectivity between the microscopic property of these systems and their macroscopic properties. The pioneering work of Sir Ludwig Boltzmann provided this missing link. According to him, all the molecules in the macro-state do not have the same energy but are actually a distribution of momentum and energy. The number of microstates can be

A. K. Dutta, *Soft Matter*, https://doi.org/10.1007/978-981-96-0624-5_2

directly related to the entropy of the system which is a measure of disorder of that system [3–5].

In thermodynamics, "everything around us" that is all objects and their surroundings constitutes the universe. An object of interest under study is defined as a 'system' and the remainder of the universe is the 'surrounding'. All interactions between the "system" and its "surroundings" may be classified into three distinct cases: (a) An "Isolated" system or a completely insulated system that does not allow mass or energy transport from the system to its surroundings or vice-versa. (b) A "Closed" system, where exchange of energy between the system and its surroundings is possible but not mass transfer, and (c) An "Open" system, where transport and exchange of matter and energy from the system to its surroundings and vice-versa is possible [4, 5].

Classical thermodynamics assumes that the system is in a 'state of thermodynamic equilibrium' which means it has attained a time-invariant state in which no further physical or chemical change will take place even over very long periods of time, ideally infinity. This implies that the system is in a state of mechanical, thermal, and chemical equilibrium [4, 5]. The simplest model systems studied in classical thermodynamics are gases at low or moderate pressures as the molecules are located far apart from each other and hence effects due to intermolecular forces are negligible which is theoretically equivalent to a system of non-interacting particles. The thermodynamic state of a gas at any instance of time is defined by three accurately measurable parameters pressure (P), volume (V), and temperature (T). All variables describing a system in thermodynamic equilibrium are classified as either (a) extensive or (b) intensive. Extensive variables are those in which properties may be added up. as the size or amount of matter or energy in the system and are additive e.g., mass (m), volume (V), number of molecules (N), internal energy (U), enthalpy (H), and entropy (S) etc. Intensive variables are independent of the size of a system but specify a local characteristic of that system, e.g., temperature (T), pressure (P), density (ρ), electric field strength (E), etc.

As described earlier, any thermodynamic state may be expressed in terms of P, V, and T. Thus if an initial state is represented by subscript I, we have S_I (P_I, V_I, T_I) and final state by subscript F that is S_F (P_F, V_F, T_F) then transition from an initial state (I) to a final state (F) may proceed through a large number of pathways that may again be executed through an infinite number of infinitesimally small steps along each pathway. Going from one state (S_I) to another (S_F) is always associated with the sum of changes (gain or loss) of energy of the system along each pathway. But, on completing a loop between (S_I) to (S_F) and back, the net energy gained or lost by the system is zero. Such a process is defined as reversible. Reversible processes are ideal cases while real processes may be both time-dependent and irreversible. In many real systems, equilibrium may shift spontaneously in one direction as in many closed systems leading to a permanent change of state that relates to an irreversible process. In this chapter, we will consider systems that are in state of thermodynamic equilibrium or infinitesimally close to it. Non-equilibrium states will not be considered and are beyond the scope of this chapter.

2.2 Fundamental Laws of Thermodynamics

Briefly, the principles of thermodynamics may be contained in three rules, the zeroth, first, and second laws of thermodynamics.

2.2.1 *The Zeroth Law of Thermodynamics*

States that if a system (A) is at thermal equilibrium with another system (B) which in turn is at thermal equilibrium with the system (C), then systems (A) and (C) are also in a state of thermal equilibrium [4].

2.2.2 *The First Law of Thermodynamics*

The First Law of Thermodynamics or the law of conservation of energy states that energy can neither be created nor destroyed but is simply converted from one form to another, In other words, change in the internal energy of any system is equal to the difference between the heat added to the system by its surroundings and work done by the system on its surroundings [5, 6] or the sum total of changes to energy of any system and its surroundings is always zero. Mathematically, it is expressed as $\mathrm{d}U = \mathrm{d}Q + \mathrm{d}W_{\mathrm{Total}}$ where change in the internal energy ($\mathrm{d}U$) is a sum total of the changes in its heat content ($\mathrm{d}Q$) and the total work done ($\mathrm{d}W_{\mathrm{Total}}$) on the system. Total work $\mathrm{d}W_{\mathrm{Total}} = \mathrm{d}W_{\mathrm{PV}} + \mathrm{d}W_{\mathrm{NPV}}$ has two components, (a) P–V work arising from expansion or compression of the gas and (b) non P–V work is given by ($\mathrm{d}W_{\mathrm{NPV}}$) arising from all other forms of work, e.g. expansion of the surface area (A) of a film by $\mathrm{d}A$ against surface tension forces γ given by $\mathrm{d}W_{\gamma} = \gamma\mathrm{d}A$ or re-orientation and re-alignment of electric or magnetic dipoles in a medium under the influence of an electric (E) or magnetic (B) field is given by $\mathrm{d}W_E = E\mathrm{d}P$ or $\mathrm{d}W_M = B\mathrm{d}M$ where P is the electric polarization and M is the magnetic field strength. Such work is referred to as non-P–V work [4–6]. Thus combining all the contributions, the first law of thermodynamics may be written as

$$\begin{aligned}\mathrm{d}U &= \mathrm{d}Q + \mathrm{d}W_{PV} + \mathrm{d}W_{NPV} \\ &= \mathrm{d}Q - P\mathrm{d}V + \gamma\mathrm{d}A - E\mathrm{d}P - B\mathrm{d}M\,.\end{aligned} \tag{2.2.2.1}$$

When

$$\begin{aligned}\mathrm{d}W_{\mathrm{NPV}} &= 0\,\text{then}\,\mathrm{d}U = \mathrm{d}Q + \mathrm{d}W_{PV} + \mathrm{d}W_{NPV} \\ &= T\mathrm{d}S - P\mathrm{d}V\,.\end{aligned} \tag{2.2.2.2}$$

2.2.3 *The Second Law of Thermodynamics*

The Second Law of Thermodynamics states that heat does not flow spontaneously from a cold body to a hot body.

2.2.4 *The Third Law of Thermodynamics*

The Third Law of Thermodynamics states that the entropy of a perfect crystal at "zero" degree Kelvin is zero.

2.3 Entropy

Entropy (S) of a system is a measure of its thermal energy per unit temperature that is unavailable for doing useful work. It is an indicator of molecular disorder or randomness of the system. Boltzmann provided the relation $S = k_B \ln \Omega$ where k_B is the Boltzmann constant and Ω is the number of microstates existing in that system. In a closed system, entropy increases and attains a maximum value at equilibrium that corresponds to a state of maximum randomness.

2.4 The Carnot Cycle

The Carnot cycle was first conceptualized by Sir Sadi Carnot in the early nineteenth century and forms the heart of classical thermodynamics. This is an idealized model of a heat engine that operates through four cycles (a) an isothermal expansion where heat is absorbed from a heat source at a temperature $T1$, (b) an adiabatic expansion that is achieved by doing work and the remaining heat energy is returned to the sink (c) an isothermal compression and (d) an adiabatic compression at lower temperature when work is received.

Figure 2.1 The P–V diagram of an ideal gas undergoing an isothermal and adiabatic expansion followed by an isothermal and adiabatic compression. Figure 2.1 shows a reversible Carnot engine operated infinitesimally slowly, where the amount of heat Q_1 extracted from a hot source and Q_2 is returned to a cold reservoir maintained at temperatures T_1 and T_2, respectively, the efficiency (η) of such a cycle is given by

$$\eta = 1 - \frac{Q_2}{Q_1} = 1 - \frac{T_2}{T_1}. \tag{2.4.1}$$

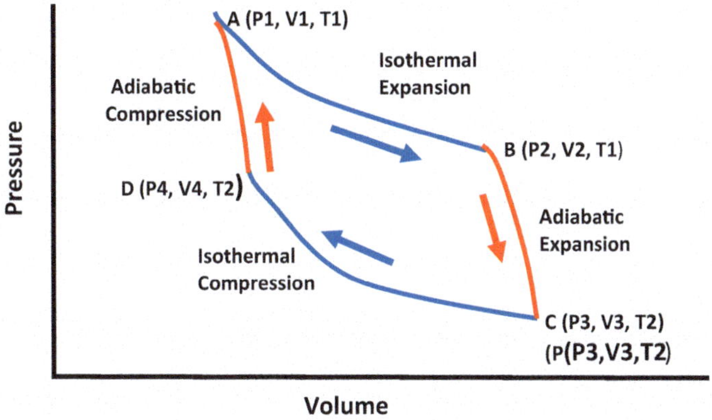

Fig. 2.1 Four cycles of an ideal Carnot engine

But $T_2 < T_1$ as T_2 is the temperature of the sink and T_1 is the temperature of the source that yields $\eta < 1$ that is true for all real systems. Again, $\eta < 1$ implies $Q_2 < Q_1$.

Hence Eqs. (2.4.1) and (2.4.2) simplifies to

$$\frac{Q_2}{Q_1} = \frac{T_2}{T_1} \quad \text{or} \quad \frac{Q_1}{T_1} = \frac{Q_2}{T_2}. \tag{2.4.2}$$

Because expansion and compression were achieved quasi-statically at an infinitesimally slow pace maintaining thermodynamic equilibrium at every instant in time and assuming no loss of energy as in an ideal isothermal process we get $T_1 \sim T_2 = T$, we have $\frac{Q_1 - Q_2}{T} = \frac{\mathrm{d}Q}{T} = \mathrm{d}S$ where S is the entropy.

In a real system, $Q_2 < Q_1$, hence $\mathrm{d}S > 0$ while for an ideal system $Q_2 = Q_1$, we get $\mathrm{d}S = 0$.

Where $T\mathrm{d}S > 0$ for real irreversible processes and $T\mathrm{d}S = 0$ for ideal reversible processes.

This is the mathematical statement of the Second law of thermodynamics.

2.5 Thermodynamic State Functions: Extensive and Intensive Parameters

The thermodynamic state of a system may be described by extensive macroscopic parameters like entropy (S), volume (V), and number of moles (n) of molecules in the system. Because state functions do not depend on how a given state was attained but on the magnitude of parameters like S, n, V, etc. of the initial and final states, the internal energy U of a system is a state function described as $U = (S, V, n_i)$ and hence an infinitesimally small change in (U) may be expressed as

$$dU = \left(\frac{\partial U}{\partial S}\right)_{V,n} dS + \left(\frac{\partial U}{\partial V}\right)_{S,n} dV + \sum_i \left(\frac{\partial U}{\partial n_i}\right)_{V,S} dn_i. \tag{2.5.1}$$

Therefore the first law of thermodynamics may be written as

$$dU = TdS - PdV + \sum_{i=1}^{N} \mu_i dn_i \tag{2.5.2}$$

where μ_i is the chemical potential of the i-th species in the system. Comparing Eqs. (2.3.1) and (2.3.2) we get $T = \left(\frac{\partial U}{\partial S}\right)_{V,n}$; $P = -\left(\frac{\partial U}{\partial V}\right)_{S,n}$ and $\mu_i = \left(\frac{\partial U}{\partial n_i}\right)_{V,S}$ implying that these three intensive parameters T, P, μ may be expressed as partial derivatives of the state function (U) with respect to the extensive parameters S, V, n and independent of the intensive parameters.

2.6 Euler and the Gibbs–Duhem Equations

The first law of thermodynamics is actually a statement of conservation of energy and mathematically stated as

$$dU = dQ + dW_{PV} + dW_{NPV} = TdS - PdV + \sum_i \mu dn_i. \tag{2.6.1}$$

Integrating, we get Euler's equation

$$U = TS - PV + \sum_i \mu_i n_i. \tag{2.6.2}$$

Note that T, P, and μ are intensive parameters, and S, V, and n are extensive parameters and each item is a product of an intensive and extensive parameter and has dimensions of energy. Differentiating both sides of the Euler Eq. (2.6.2) yields

$$dU = TdS + SdT - PdV - VdP + \sum_i \mu_i dn_i + \sum_i n_i d\mu_i. \tag{2.6.3}$$

Comparing Eqs. (2.6.1) and (2.6.3) we get the Gibbs–Duhem equation [6–8]

$$SdT - VdP + \sum_i n_i d\mu_i = 0. \tag{2.6.4}$$

Note that the integrable variables (T, P, μ) in the Gibbs–Duhem equation are intensive variables and are conjugates to the extensive variables (S, V, n). Considering isothermal and isobaric conditions, we get $\sum_i n_i d\mu_i = 0$. Thus for a two-component mixture $\mathrm{d}\mu_1 = -\frac{n_2}{n_1}\mathrm{d}\mu_2$ which implies that the chemical potentials of the species in a mixture are not independent of each other but depend on the relative concentrations of the species [7, 8].

2.7 Enthalpy (*H*)

As mentioned, the internal energy of a system U is a state function, and small change in internal energy $\mathrm{d}U$ is given by $\mathrm{d}U = \mathrm{d}Q + \mathrm{d}W_{PV} + \mathrm{d}W_{NPV} = \mathrm{d}Q - P\mathrm{d}V + \sum_i \mu_i \mathrm{d}n_i$ assuming only P–V work and no exchange of particles as in a closed system we get $\mathrm{d}Q = \mathrm{d}U - \mathrm{d}W = \mathrm{d}U - P\mathrm{d}V$ which on integration yields $\int_{Q1}^{Q2} \mathrm{d}Q = \int_{U1}^{U2} \mathrm{d}U - \int_{V1}^{V2} P\mathrm{d}V$ the equation

$$Q_2 - Q_1 = (U_2 - U_1) + P(V_2 - V_1) \quad \text{or} \quad \Delta Q = \Delta U + P\Delta V$$

The right-hand side may be expressed as $\int_{H1}^{H2} \mathrm{d}H = \int_1^2 \mathrm{d}(U + PV)$ where H is the enthalpy and is equal to Q when P–V work is zero. Integrating, we have

$$H = U + PV. \tag{2.7.1}$$

Differentiating Eq. (2.7.1) yields $\mathrm{d}H = T\mathrm{d}S + V\mathrm{d}P$ hence

$$H = H(S, P). \tag{2.7.2}$$

It follows immediately that at constant pressure

$$\mathrm{d}H = T\mathrm{d}S = \mathrm{d}Q. \tag{2.7.3}$$

Thus at constant pressure, change in enthalpy is equal to the change in the quantity of heat received or lost by the system.

2.8 Helmholtz (*F*) Free Energy and Gibbs (*G*) Free Energy

The entropy concept allows defining two new thermodynamic variables namely (*a*) Helmholtz free energy (F) and the (*b*) Gibbs free energy (G). While the Helmholtz free energy (F) corresponds to the maximum amount of work that is obtainable from a system at constant temperature and volume. (2) Gibbs free energy (G) corresponds to the maximum amount of work a system can do at constant temperature and pressure [7, 8].

The expression for Helmholtz free energy (F) is given by $F = U - TS$ which in differential form is

$$\mathrm{d}F = T\mathrm{d}S - p\mathrm{d}V + \Sigma\mu_i \mathrm{d}n_i - T\mathrm{d}S - S\mathrm{d}T = -p\mathrm{d}V - S\mathrm{d}T. \tag{2.8.1}$$

Hence

$$F = F(V, T). \tag{2.8.2}$$

Assuming no exchange of particles as in a closed system $\mathrm{d}n_i = 0$ and under isothermal conditions $\mathrm{d}T = 0$ we get $\mathrm{d}F = -p\mathrm{d}V = \mathrm{d}W_{\mathrm{rev}}$ implying that the Helmholtz free energy (F) corresponds to the P–V work done in a reversible process and hence corresponds to the maximum work that may be done by the system as maximum work achievable in a reversible process at a constant temperature.

The Gibbs free energy (G) is given by $G = H - TS = U + pV - TS$ the differential form of which is

$$\begin{aligned} \mathrm{d}G = {} & T\mathrm{d}S - p\mathrm{d}V + p\mathrm{d}V + V\mathrm{d}p \\ & + \sum_i \mu_i \mathrm{d}n_i - T\mathrm{d}S - S\mathrm{d}T = V\mathrm{d}p - S\mathrm{d}T. \end{aligned} \tag{2.8.3}$$

Hence

$$G = G(P, T). \tag{2.8.4}$$

Considering no exchange of particles as in a closed system $\mathrm{d}n_i = 0$ and under isothermal $\mathrm{d}T = 0$ and isobaric conditions $\mathrm{d}P = 0$ we get $\mathrm{d}G = 0$ implying maximum non-PV work done by the system.

2.9 Dependence of Gibbs and Helmholtz Free Energy on Temperature: Gibbs–Helmholtz Equations

The expression for Gibbs (G) potential is $G = H - TS$ the differential form of which is given by which on differentiating with respect to T yields

$$\begin{aligned} \frac{\mathrm{d}}{\mathrm{d}T}\left(\frac{G}{T}\right) &= \frac{\mathrm{d}}{\mathrm{d}T}\left(\frac{H}{T}\right) - \frac{\mathrm{d}S}{\mathrm{d}T} \quad \text{or} \\ \frac{\mathrm{d}}{\mathrm{d}T}\left(\frac{G}{T}\right) &= \frac{1}{T}\frac{\mathrm{d}H}{\mathrm{d}T} - \frac{H}{T^2} - \frac{\mathrm{d}S}{\mathrm{d}T}. \end{aligned} \tag{2.9.1}$$

But $H = U + PV$ or $\mathrm{d}H = \mathrm{d}U + \mathrm{d}(PV) = \mathrm{d}q - p\mathrm{d}V + p\mathrm{d}V + V\mathrm{d}p = T\mathrm{d}S + V\mathrm{d}p$. Dividing by dT we have $\frac{\mathrm{d}H}{\mathrm{d}T} = T\frac{\mathrm{d}S}{\mathrm{d}T} + V\frac{\mathrm{d}p}{\mathrm{d}T}$ or $\frac{1}{T}\frac{\mathrm{d}H}{\mathrm{d}T} - \frac{\mathrm{d}S}{\mathrm{d}T} = \frac{V}{T}\frac{\mathrm{d}p}{\mathrm{d}T}$ and hence at constant pressure $\frac{1}{T}\frac{\mathrm{d}H}{\mathrm{d}T} - \frac{\mathrm{d}S}{\mathrm{d}T} = 0$.

Therefore

$$\frac{\mathrm{d}}{\mathrm{d}T}\left(\frac{G}{T}\right) = \frac{1}{T}\frac{\mathrm{d}H}{\mathrm{d}T} - \frac{H}{T^2} - \frac{\mathrm{d}S}{\mathrm{d}T}. \tag{2.9.2}$$

At constant pressure, the above equation modifies to

$$\left(\frac{\mathrm{d}}{\mathrm{d}T}\left(\frac{G}{T}\right)\right)_p = -\frac{H}{T^2}. \tag{2.9.3}$$

Similarly, starting from

$$F = U - TS \text{ and } \mathrm{d}F = \mathrm{d}U - \mathrm{d}(TS) = -p\mathrm{d}V - S\mathrm{d}T. \tag{2.9.4}$$

Differentiating w.r.t. to T yields

$$\frac{\mathrm{d}}{\mathrm{d}T}\left(\frac{F}{T}\right) = \frac{1}{T}\frac{\mathrm{d}U}{\mathrm{d}T} - \frac{U}{T^2} - \frac{\mathrm{d}S}{\mathrm{d}T}. \tag{2.9.5}$$

At constant volume, the above modifies to

$$\left(\frac{\mathrm{d}}{\mathrm{d}T}\left(\frac{F}{T}\right)\right)_v = -\frac{U}{T^2}. \tag{2.9.6}$$

2.10 Important Identities in Thermodynamics

The state of any closed system may be described by at least two parameters out of the four most commonly used parameters (P, T, S, V) used to represent a thermodynamic system. While P and T are intensive parameters, S and V are extensive parameters. The four thermodynamic quantities frequently used are U, F, G, and H where

$$U = U(S, V);\ F = F(T, V);\ H = H(S, P) \text{ and } G = G(T, P).$$

Since $U = U(S, V)$ we have

$$\mathrm{d}U = \left(\frac{\partial U}{\partial S}\right)_V \mathrm{d}S + \left(\frac{\partial U}{\partial V}\right)_S \mathrm{d}V \tag{2.10.1}$$

and for a reversible reaction $\mathrm{d}U = T\mathrm{d}S - P\mathrm{d}V$ therefore we have:

$$T = \left(\frac{\partial U}{\partial S}\right)_V \text{ and } P = -\left(\frac{\partial U}{\partial V}\right)_S. \tag{2.10.2}$$

Similarly, $F = F(T, V)$ we have $\mathrm{d}F = \left(\frac{\partial F}{\partial T}\right)_V \mathrm{d}T + \left(\frac{\partial F}{\partial V}\right)_T \mathrm{d}V$.

Again as $\mathrm{d}F = -S\mathrm{d}T - P\mathrm{d}V$ we get:

$$S = -\left(\frac{\partial F}{\partial T}\right)_V \quad P = -\left(\frac{\partial F}{\partial V}\right)_T. \tag{2.10.3}$$

Similarly, Enthalpy (H) has been described as $H = H(S, P)$. Therefore, $\mathrm{d}H = \left(\frac{\partial H}{\partial S}\right)_p \mathrm{d}S + \left(\frac{\partial H}{\partial P}\right)_S \mathrm{d}P$.

Given that $\mathrm{d}H = T\mathrm{d}S + V\mathrm{d}p$ we have:

$$T = \left(\frac{\partial H}{\partial S}\right)_p \text{ and } V = \left(\frac{\partial H}{\partial P}\right)_S. \tag{2.10.4}$$

And Gibbs free energy (G) can be described as

$$G = G(T, P)$$

and $\mathrm{d}G = \left(\frac{\partial G}{\partial T}\right)_V \mathrm{d}T + \left(\frac{\partial G}{\partial P}\right)_T \mathrm{d}P$. Again as $\mathrm{d}G = -S\mathrm{d}T + V\mathrm{d}P$ it follows:

$$S = -\left(\frac{\partial G}{\partial T}\right)_V \quad V = \left(\frac{\partial G}{\partial p}\right)_T. \tag{2.10.5}$$

Therefore, we have the following identities;

$$T = \left(\frac{\partial U}{\partial S}\right)_V = \left(\frac{\partial H}{\partial S}\right)_p \tag{2.10.6}$$

$$P = -\left(\frac{\partial U}{\partial V}\right)_S = -\left(\frac{\partial F}{\partial V}\right)_T \tag{2.10.7}$$

$$V = \left(\frac{\partial H}{\partial P}\right)_S = \left(\frac{\partial G}{\partial P}\right)_T \tag{2.10.8}$$

$$S = -\left(\frac{\partial F}{\partial T}\right)_V = -\left(\frac{\partial G}{\partial T}\right)_p. \tag{2.10.9}$$

2.11 Maxwell's Thermodynamic Relations

The internal energy for a closed system is given by $U = U(S, V)$.

Then $\mathrm{d}U = \left(\frac{\partial U}{\partial S}\right)_V \mathrm{d}S + \left(\frac{\partial U}{\partial V}\right)_S \mathrm{d}V$ where $T = \left(\frac{\partial U}{\partial S}\right)_V$; $-P = \left(\frac{\partial U}{\partial V}\right)_S$.

As $\mathrm{d}U = T\mathrm{d}S - p\mathrm{d}V$ the partial derivatives $\left(\frac{\partial T}{\partial V}\right)_S = \frac{\partial}{\partial V}\left[\left(\frac{\partial U}{\partial S}\right)_V\right]_S$ and $-\left(\frac{\partial p}{\partial S}\right)_V = \frac{\partial}{\partial S}\left[\left(\frac{\partial U}{\partial V}\right)_S\right]_V$.

And $\mathrm{d}U = T\mathrm{d}S - p\mathrm{d}V$ gives $\left(\frac{\partial T}{\partial V}\right)_S = -\left(\frac{\partial P}{\partial S}\right)_V$.
Similarly, $\mathrm{d}F = -P\mathrm{d}V - S\mathrm{d}T$ we get $\left(\frac{\partial P}{\partial T}\right)_V = \left(\frac{\partial S}{\partial V}\right)_p$.
$\mathrm{d}G = V\mathrm{d}P - S\mathrm{d}T$ we get $\left(\frac{\partial V}{\partial T}\right)_p = -\left(\frac{\partial S}{\partial P}\right)_T$.
$\mathrm{d}H = T\mathrm{d}S + V\mathrm{d}P$ we get $\left(\frac{\partial T}{\partial P}\right)_S = \left(\frac{\partial V}{\partial S}\right)_p$.
This set of four equations is called Maxwell's equations and connects the variation of one pair of parameters S, V, P, and T with another pair [4–8].

2.12 Specific Heat Parameters C_P and C_V

It follows from the first law that $\mathrm{d}Q = \mathrm{d}U + \mathrm{d}W_{PV} + \mathrm{d}W_{NPV} = T\mathrm{d}S - P\mathrm{d}V + \mathrm{d}W_{NPV} = T\mathrm{d}S$ when no non-PV work is done and volume of the system remains constant. Dividing both sides with respect to $\mathrm{d}T$ we have

$$\frac{\mathrm{d}Q}{\mathrm{d}T} = T\frac{\mathrm{d}S}{\mathrm{d}T} \text{ or } C_V = T\left(\frac{\mathrm{d}S}{\mathrm{d}T}\right)_V = \left(\frac{\mathrm{d}U}{\mathrm{d}T}\right)_V = \left(\frac{\mathrm{d}Q}{\mathrm{d}T}\right)_V. \tag{2.12.1}$$

Again $\mathrm{d}H = \mathrm{d}U + \mathrm{d}(PV) = T\mathrm{d}S + V\mathrm{d}P$.
Dividing both sides with respect to $\mathrm{d}T$ we have

$$\frac{\mathrm{d}H}{\mathrm{d}T} = T\frac{\mathrm{d}S}{\mathrm{d}T} + V\left(\frac{\mathrm{d}P}{\mathrm{d}T}\right) = T\left(\frac{\mathrm{d}S}{\mathrm{d}T}\right).$$

Which at constant pressure simplifies to

$$C_P = T\left(\frac{\mathrm{d}S}{\mathrm{d}T}\right)_P = \left(\frac{\mathrm{d}H}{\mathrm{d}T}\right)_P. \tag{2.12.2}$$

The thermal expansion coefficient (α) is defined as $\alpha_P = \frac{1}{V}\left(\frac{\mathrm{d}V}{\mathrm{d}T}\right)_P = \left(\frac{\mathrm{d}V/V}{\mathrm{d}T}\right)_P = \frac{\mathrm{d}(\ln V)_P}{\mathrm{d}T}$ and isothermal compressibility coefficient (β) is defined as $\beta_T = -\frac{1}{V}\left(\frac{\mathrm{d}V}{\mathrm{d}P}\right)_T = -\left(\frac{\mathrm{d}V/V}{\mathrm{d}P}\right)_T = -\frac{\mathrm{d}(\ln V)_T}{\mathrm{d}P}$.
It may be readily shown that $C_P = C_V + \frac{\alpha_P^2 VT}{\beta_T}$ that confirms

$$C_P > C_V. \tag{2.12.3}$$

Similarly, adiabatic compressibility coefficient (β_S) is given as

$$\beta_S = -\frac{1}{V}\left(\frac{\mathrm{d}V}{\mathrm{d}P}\right)_S = -\left(\frac{\mathrm{d}V/V}{\mathrm{d}P}\right)_S = -\frac{\mathrm{d}(\ln V)_S}{\mathrm{d}P}$$

and it readily follows that

$$\frac{\beta_T}{\beta_S} = \frac{C_P}{C_V} \tag{2.12.4}$$

that demonstrates adiabatic compressibility (β_S) is always less than isothermal compressibility (β_T) [6–9].

2.13 Thermodynamics of Open Systems

By definition, open systems allow exchange of matter and energy with its surroundings. Therefore, $U = U(S, V, n)$ hence

$$\mathrm{d}U = \left(\frac{\partial U}{\partial S}\right)_{V,n}\mathrm{d}S + \left(\frac{\partial U}{\partial V}\right)_{S,n}\mathrm{d}V + \left(\frac{\partial U}{\partial n}\right)_{V,S}\mathrm{d}n. \tag{2.13.1}$$

If μ_i be the chemical potential for the molecular species "*i*", then the expression for internal energy can be written as

$$\mathrm{d}U = T\mathrm{d}S - P\mathrm{d}V + \sum_i \mu_i \mathrm{d}n_i. \tag{2.13.2}$$

Comparing the above two equations we have the following:

$$T = \left(\frac{\partial U}{\partial S}\right)_{T,n} P = -\left(\frac{\partial U}{\partial V}\right)_{S,n} \text{ and } \mu_i = \left(\frac{\partial U}{\partial n_i}\right)_{V,S}. \tag{2.13.3}$$

The fundamental thermodynamic potentials are $H = H(S, P, n)$, $F = F(T, V, n)$,$G = G(T, P, n)$ and

$$\mathrm{d}H = T\mathrm{d}S + V\mathrm{d}P + \sum_i \mu_i \mathrm{d}n_i \text{ and}$$

$$\mathrm{d}H = \left(\frac{\partial H}{\partial S}\right)_{p,n}\mathrm{d}S + \left(\frac{\partial H}{\partial P}\right)_{S,n}\mathrm{d}P + \left(\frac{\partial H}{\partial n}\right)_{S,p}\mathrm{d}n$$

$$T = \left(\frac{\partial H}{\partial S}\right)_{P,n} V = \left(\frac{\partial H}{\partial P}\right)_{S,n} \text{ and } \mu_i = \left(\frac{\partial H}{\partial n_i}\right)_{S,P}. \tag{2.13.4}$$

Similarly, $\mathrm{d}F = -S\mathrm{d}T - P\mathrm{d}V + \sum_i \mu_i \mathrm{d}n_i$ and $\mathrm{d}F = \left(\frac{\partial F}{\partial T}\right)_{V,n}\mathrm{d}T + \left(\frac{\partial F}{\partial V}\right)_{T,n}\mathrm{d}V + \left(\frac{\partial F}{\partial n}\right)_{T,V}\mathrm{d}n$. Then

$$S = -\left(\frac{\partial F}{\partial T}\right)_{V,n} P = -\left(\frac{\partial F}{\partial V}\right)_{T,n} \text{ and } \mu_i = \left(\frac{\partial F}{\partial n}\right)_{T,V}. \tag{2.13.5}$$

Again $dG = -SdT + VdP + \sum_i \mu_i dn_i$ and

$$dG = \left(\frac{\partial G}{\partial T}\right)_{p,n} dT + \left(\frac{\partial G}{\partial P}\right)_{T,n} dP + \left(\frac{\partial G}{\partial n}\right)_{T,p} dn \tag{2.13.6}$$

$$S = -\left(\frac{\partial G}{\partial T}\right)_{P,n} V = \left(\frac{\partial G}{\partial P}\right)_{T,n} \text{ and } \mu_i = \left(\frac{\partial G}{\partial n_i}\right)_{T,P}. \tag{2.13.7}$$

Summing up we have

$$T = \left(\frac{\partial H}{\partial S}\right)_{P,n} V = \left(\frac{\partial H}{\partial P}\right)_{S,n} = \left(\frac{\partial G}{\partial P}\right)_{T,n} S = -\left(\frac{\partial G}{\partial T}\right)_{P,n} \text{ and}$$

$$\mu_i = \left(\frac{\partial G}{\partial n_i}\right)_{T,P} = \left(\frac{\partial H}{\partial n_i}\right)_{S,P} = \left(\frac{\partial F}{\partial n_i}\right)_{T,V} = \left(\frac{\partial U}{\partial n_i}\right)_{V,S}.$$

Thus chemical potential μ_i may be expressed as a partial derivative of one of the energy terms U, H, F, G with the number of particles (n_i) in an open system.

2.14 Phase Transitions

It is well known that on heating or cooling most materials undergo change in its volume, density, and often appearance. Water is an excellent example in this context that on cooling water which is a liquid at an ambient temperature of 25 °C gets transformed to ice at 0 °C. Heating water causes its volume to expand and at 100 °C it gets converted to steam. Similar behavior is observed for most materials. Figure 2.2 shows the pressure (P) versus Temperature (T) profile for a material. Along each curve, at least two phases coexist at equilibrium while all three phases coexist at the triple point at equilibrium. Beyond the curves, a single phase exists as clearly shown in Fig. 2.2. We represent the phases as 1, 2, and 3 where each numerical can be assigned any phase solid, liquid, and gas. A solid-to-gas transition is sublimation, a solid-to-liquid is melting, and liquid-to-vapor is vaporization as shown by arrows in Fig. 2.2.

Let us consider a situation where any two phases (1 and 2 say) exist at equilibrium. This means that the system is in a state of mechanical, thermal, and energy equilibrium. Then $\mu_1(P, T) = \mu_2(P, T)$.

Considering Gibbs free energy we have

$$dG = VdP - SdT + \sum_i \mu_i dN_i$$

or

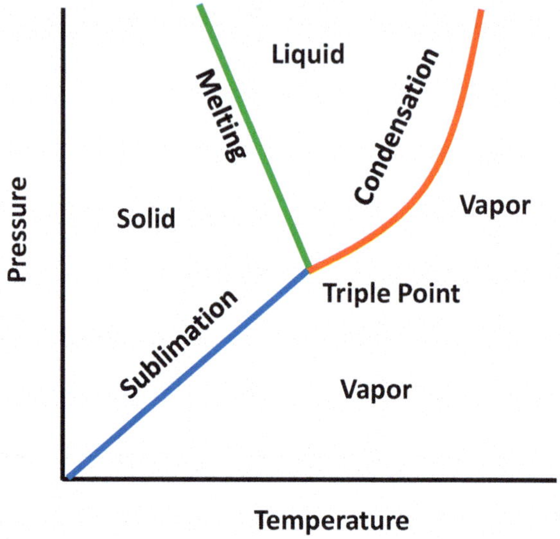

Fig. 2.2 Schematic of the phase diagram (P vs. T)

$$\frac{\mathrm{d}G}{\mathrm{d}T} = V\frac{\mathrm{d}P}{\mathrm{d}T} - S \tag{2.14.1}$$

assuming a closed system, that is $\sum_i \mu_i \mathrm{d}N_i = 0$.

Therefore, for phases 1 and 2 at equilibrium is

$$\frac{\mathrm{d}G_1}{\mathrm{d}T} = V_1\frac{\mathrm{d}P_1}{\mathrm{d}T} - S_1 \tag{2.14.2}$$

and

$$\mathrm{d}T. \tag{2.14.3}$$

Subtracting and simplifying we get

$$\frac{\mathrm{d}(G_1 - G_2)}{\mathrm{d}T} = V_1\frac{\mathrm{d}P_1}{\mathrm{d}T} - V_2\frac{\mathrm{d}P_2}{\mathrm{d}T} - S_1 + S_2. \tag{2.14.4}$$

At the phase boundary.

$P_1 = P_2 = P(\text{say})$ and $G_1 = G_2$.that yields

$$\frac{\mathrm{d}P}{\mathrm{d}T} = \frac{S_1 - S_2}{V_1 - V_2} \text{ as } G_1 = G_2. \tag{2.14.5}$$

Or

$$\frac{\mathrm{d}P}{\mathrm{d}T} = \frac{S_1 - S_2}{V_2 - V_1} = \frac{\Delta H}{T_m \Delta V} \tag{2.14.6}$$

where enthalpy changes due to phase transition is $\Delta H = T_m \Delta S$.

This equation is the Clausius–Clapeyron equation and the enthalpy term ΔH corresponds to the enthalpy change occurring at the phase boundaries. In case of a liquid–vapor transition ΔH corresponds to the latent heat of vaporization while for a solid–liquid transition ΔH is the latent heat of fusion.

It follows from Eq. (2.14.1) that Gibbs free energy is $\frac{dG}{dT} = V\frac{dP}{dT} - S$.

Hence at constant pressure, we get

$$\left|\frac{dG_s}{dT}\right|_p = -S_s; \left|\frac{dG_l}{dT}\right|_p = -S_l \text{ and } \left|\frac{dG_v}{dT}\right|_p = -S_v. \tag{2.14.7}$$

Suffixes correspond to solid (s), Liquid (l) and gas (g) phases.

A close look at Fig. 2.3 shows that $S_v > S_l > S_s$ that is consistent with the fact that molecules in vapor phase correspond to the highest state of positional and orientation disorder and lowest in case of solids. The very fact that the entropies are different for the different phases suggests that the first order derivative of the Gibbs free energy versus Temperature plot are discontinuous for volume (V), enthalpy (H), Entropy (S) as well as for coefficient of thermal expansion (α), isothermal compressibility (β) and specific heat at constant pressure (C_p) as shown in Fig. 4a, b.

This feature indicates that the first-order transitions are discontinuous at the phase transition temperatures T_{mp} and T_{bp} correspond to the melting point and boiling point at the solid–liquid and liquid–vapor boundaries as shown in the schematic. The second derivative $\left|\frac{d^2G_s}{dT^2}\right|_p = -\frac{dS_s}{dT}$ and $\left|\frac{d^2G_l}{dT^2}\right|_p = -\frac{dS_l}{dT}$ but at the transition temperature T_{mp} we have $S_s = S_l$. Hence the second-order derivative of G is continuous. Similarly, it may be shown that Enthalpy, Volume, and Entropy α, β, and C_p are discontinuous. It may be pointed out that phase transitions in glasses are second-order phase transitions [9, 10].

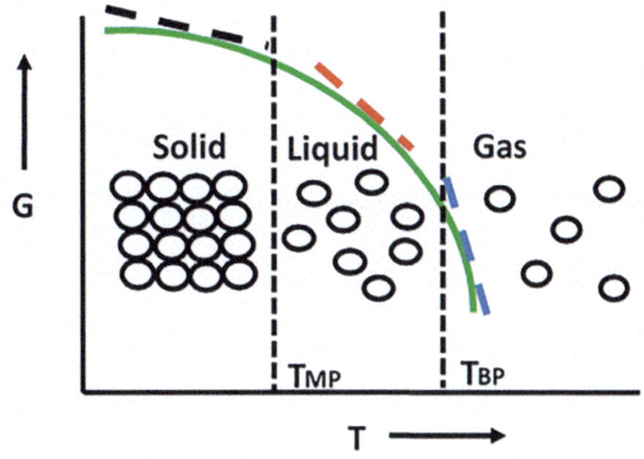

Fig. 2.3 Schematic of Gibbs free energy versus temperature. T_{MP} and T_{BP} correspond to melting point (T_{MP}) and boiling point (T_{BP}) temperatures

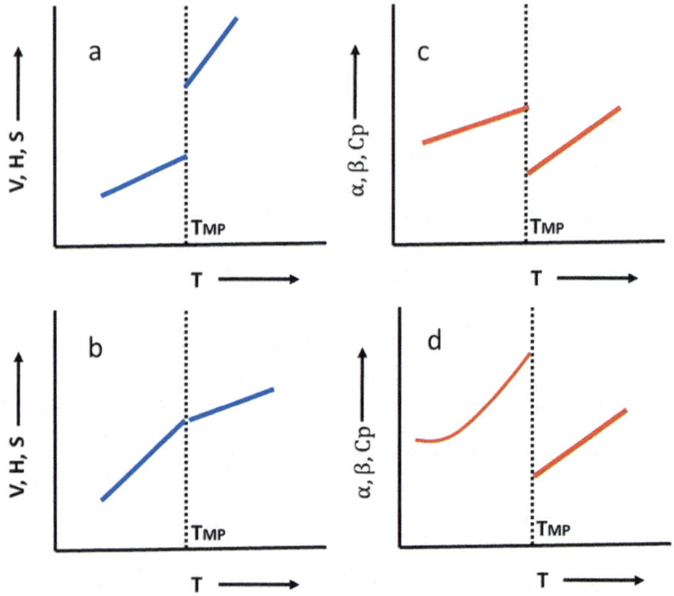

Fig. 2.4 Schematic of changes in volume (V), enthalpy (H), and entropy (S) as a function of Temperature for **a** first-order transition and **b** second-order transition. Schematic of changes in thermal expansion coefficient (α), isothermal compressibility (β), and specific heat capacity (C_p) at constant pressure as a function of Temperature for **c** first-order transition and **d** second-order transition

2.15 Homogeneous Nucleation

In homogeneous nucleation, nuclei of the new phase spontaneously appear in the absence of foreign particles that act as seeds for the new phase. Considering a spherical droplet of radius r is formed, we can calculate the change in Gibbs free energy as

$$G_{\text{Total}} = G_s + G_v \tag{2.15.1}$$

where the surface excess energy is $G_s = 4\pi r^2 \gamma$ and volume excess energy is $G_v = \frac{4}{3}\pi r^3 G_v$.

Thus total

$$G_{\text{Total}} = 4\pi r^2 G_S + \frac{4}{3}\pi r^3 G_V. \tag{2.15.2}$$

Differentiating with respect to r is

$$\frac{\mathrm{d}G_{\text{Total}}}{\mathrm{d}r} = 8\pi r G_S + 4\pi r^2 G_v. \tag{2.15.3}$$

Putting $\frac{dG_{\text{Total}}}{dr} = 0$, yields

$$r = r_c = -2\frac{G_S}{G_V}. \tag{2.15.4}$$

And

$$G_c = \frac{16\pi\gamma^3}{3(G_v)^2} \tag{2.15.5}$$

where G_C corresponds to the Gibbs free energy at $r = r_c$.

As the driving force for solidification is given by

$$G_v = H_f\left(1 - \frac{T}{T_{\text{mp}}}\right). \tag{2.15.6}$$

It follows

$$G_c = -\frac{16\pi\gamma^3 T_{\text{mp}}^2}{3H_f^2(T_{\text{mp}} - T)^2}. \tag{2.15.7}$$

It must be noted that G_c is always negative (Fig. 2.5).

In summary, spontaneous nucleation will occur when $r \geq r_c$ is achieved and the activation energy for its formation is G_c. With increasing radius of the nuclei, Gibbs free energy decreases rapidly suggesting that aggregation improves the stability of the system [9, 10].

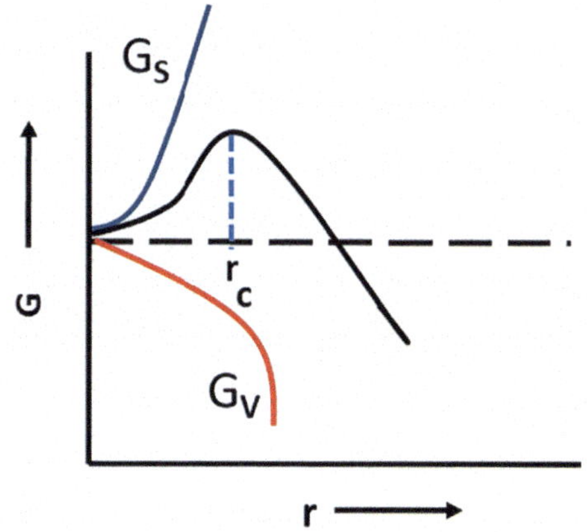

Fig. 2.5 Schematic of Gibbs free energy (G) versus the radius of the nuclei

2.16 Heterogeneous Nucleation

In heterogeneous nucleation, nuclei of the new phase appear in the presence of foreign particles that may act as seeds for the new phase. Considering a spherical droplet of radius r is formed, we can calculate the change in Gibbs free energy as

$$G_{\text{Total}} = G_s + G_v \tag{2.16.1}$$

where the surface excess energy is

$$G_s = 4\pi r^2(\gamma_{SL} - \gamma_{SV}) \tag{2.16.2}$$

and volume excess energy is $G_v = \frac{4}{3}\pi r^3$.

Here

$$\gamma_{SA} = \gamma_{SL} + \gamma_{LV}\text{Cos}\theta. \tag{2.16.3}$$

In this case, the total surface excess is given by

$$\begin{aligned} dG_s &= \pi r^2(\gamma_{SL} - \gamma_{SV}) + 2\pi r^2(1 - \cos\theta)\gamma_{LV} \\ &= 2\pi r^2(1 - \cos\theta)\gamma_{LV} + \pi r^2\gamma_{LV}\sin^2\theta. \end{aligned} \tag{2.16.4}$$

Similarly, the Volume of the droplet

$$V = \frac{1}{3}\pi r^3(1 - \cos\theta)^2(2 + \cos\theta). \tag{2.16.5}$$

On simplifying it turns out that

$$r_C = -\frac{2\gamma}{\Delta G_V} \tag{2.16.6}$$

and

$$\Delta G_C = \frac{16\pi\gamma^3}{3\Delta G_V^2}f(\theta) \tag{2.16.7}$$

where

$$f(\theta) = \frac{1}{4}(2 + \cos\theta)(1 - \cos\theta)^2. \tag{2.16.8}$$

Similar to homogeneous nucleation, here $r_C = -\frac{2\gamma}{\Delta G_V}$ suggests the existence of a critical size of nuclei that gets generated and grows bigger to form large crystals. Interestingly, the result is dependent on $f(\theta)$ where θ is the contact angle of the droplet

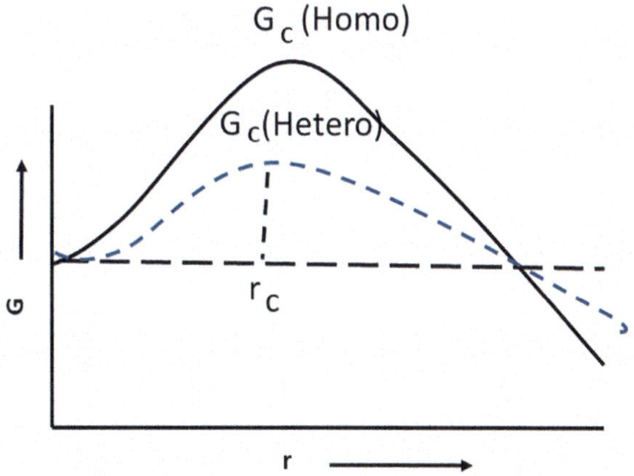

Fig. 2.6 Schematic diagram of comparison of Gibbs free energy for homogeneous and heterogeneous nucleation as a function of the nucleation radius (r)

on the foreign particle on which the seeding occurs. Comparing this result with that of homogeneous nucleation we get $G_c^{\text{Hetero}} = f(\theta)G_c^{\text{Homo}}$ implying activation energy for the heterogeneous nucleation is less than that for homogeneous nucleation as $f(\theta)$ has value between 0 and 0.5 as shown in Fig. 2.6.

2.17 Crystal Growth

Over the years, several theories related to crystal growth have emerged of which the classical nucleation theory (CNT) by LaMer and Dinegar [11] is the most popular despite several drawbacks as discussed in a later section of this chapter. Developed in the 1950s, the basic assumptions of CNT are the following:

(1) A burst of nuclei is generated within the system spontaneously and
(2) These nuclei grow over time through a diffusion-limited process to form crystals.

The homogeneous nucleation rate (J) is mathematically expressed as

$$J = K \exp\left(-\frac{\Delta G_c}{k_B T}\right) \tag{2.17.1}$$

where K is a constant, k_B is the Boltzmann constant, and ΔG_c is the Gibbs free energy of cluster formation with critical radius r_c. A close look at the expression for the Gibbs free energy of homogeneous nucleation is given as

$$\Delta G = -n\Delta\mu + \sigma A = -\frac{4}{3\nu}r^3\Delta\mu + 4\pi r^2\sigma \tag{2.17.2}$$

where r is the radius of the nuclei, ν is the molar volume, σ is the interfacial tension, and $\Delta\mu$ is the change in the chemical potential, readily shows that the first term is negative while the second term is positive and both are dependent on the single variable r. Differentiating both sides with respect to r and equating it to zero yields $r_c = \frac{2\sigma}{\Delta\mu}$. Thus at $r < r_c$, the nuclei will re-dissolve in the saturated solution to attain stable equilibrium while for particles with $r > r_c$ will grow to minimize ΔG.

References

1. Muller I (2007) A history of thermodynamics: doctrine of energy and entropy. Springer, NY
2. Fermi E (1956) Thermodynamics. Dover, N.Y.
3. Berry SR (2019) Three laws of nature. Yale University Press, New Haven & London
4. Kondepudi D (2008) Introduction to modern thermodynamics. Wiley & Sons, N.Y.
5. Luscombe JH (2018) Thermodynamics. CRC Press, Boca-Raton
6. Honig JM (2021) Thermodynamics: principles characterizing physical and chemical processes, 5th edn. Academic Press, San Deigo
7. Sherwood D, Dalby P (2018) Thermodynamics for chemists and biochemists. Oxford University Press, Oxford
8. Erbil HY (2006) Surface chemistry of solid and liquid interface. Wiley, N.Y.
9. Saggion A, Faraldo R, Pierno M (2019) Thermodynamics: fundamental principles and applications. Springer, N.Y.
10. Meier GH (2014) Thermodynamics of surfaces and interfaces: concepts in inorganic materials. Cambridge University Press, Cambridge
11. LaMer VK, Dinegar RH (1950) J Am Chem Soc 72:4847–4854

Chapter 3
Interfacial Physics

3.1 Introduction

In real systems, we often come across all three states of matter solids, liquids and gases coexisting. These may be completely different materials or be different states of the same material. The boundary separating one from the other is defined as an interface and it readily follows that four distinct phases, a solid–liquid, liquid–liquid, solid–gas and liquid–gas exist. We readily encounter such interfaces in regular day to day use products, e.g. milk which consists of tiny droplets of fat suspended in water, paint that consists of particles of pigments suspended in an oil or water base, smoke consisting of solid particles suspended in air or aerosols consisting of liquid droplets in a propellant gas.

The role and importance of an interface is readily understood if we consider two immiscible fluids *A* shown in blue circles and fluid *B* shown in red circles separated by an interface (black dotted line) as shown in Fig. 3.1. If we consider molecules *P* and *P*' located in the bulk of *A* (blue circles) and *B* (red circles) then the forces acting between *P* and surrounding *A* is zero as equal and opposite forces nullify each other as shown in the inset. Similarly, the net force on *P*' located in the bulk of *B* is also zero.

However, for *P* located at the interface of *A* and *B*, a net force acts on *P* as unequal forces come into play because forces between *P* and *A* and those between *P* and *B* are unequal. This force acts tangentially at the interface and is known as the surface tension γ_{AB}.

Colloidal systems are complex consisting of solid particles and liquid droplets dispersed in another immiscible liquid along with surfactants and excipients that are ubiquitous in food, paint, cosmetics, and pharmaceutical products among others. Complexity arises from the fact that various solid and liquid materials with surfactant polymers or their mixtures as mentioned above are present and interaction of one component with all others makes it a formidable mathematical problem. Moreover, the intermolecular forces are concentration and temperature-dependent. Another

A. K. Dutta, *Soft Matter*, https://doi.org/10.1007/978-981-96-0624-5_3

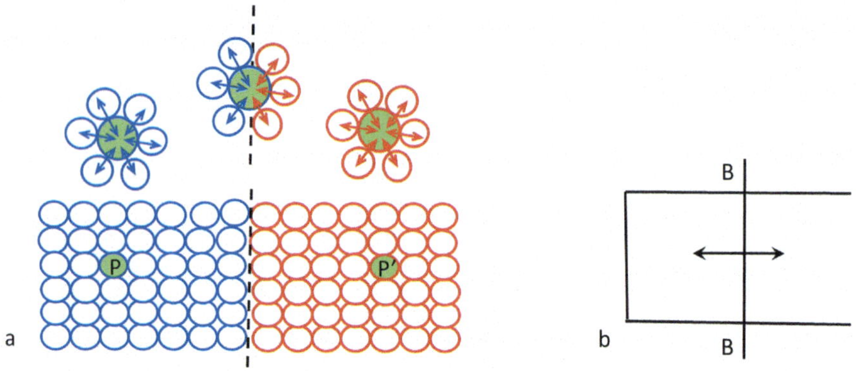

Fig. 3.1 **a** Schematic of two phases *A* (blue circles) and *B* (red circles) separated by a hypothetical interface (black dashed line). Insets show the forces acting on molecule *P* when in bulk and at the interface. **b** Model system of a rectangular trough holing a liquid of surface tension (γ) and a movable barrier BB

inherent concern with such systems are its propensity to form aggregates and phase separate out. Stabilizing such systems is a challenge as various interactions contribute toward its stability.

In this chapter, we adopt a classical thermodynamic approach to explore the role of various forces that act at the solid–liquid, liquid–vapor, and solid–vapor interfaces that determines wettability and spread of one material on another that has a key role in the design of products various products. It is anticipated that our readers especially new to this area of research will find this chapter useful in understanding the basics of the various interactions that directly contributes to product properties irrespective of its application domain whether it be detergents, food or adhesives.

3.2 Thermodynamics of Interfaces

An interface is a plane that separates two material continua (Fig. 3.1a) for example a liquid surface in contact with its vapor or two immiscible liquids in contact with each other or a liquid drop on a solid substrate. Throughout the chapter we will assume the interface to have infinitesimal thickness and both exchange of molecules as well as energy across the interface is allowed, i.e., these systems are open systems unless specifically mentioned. If we consider a trough containing a liquid at equilibrium with air (say) and the surface area of the trough may be altered by moving a barrier across the surface of the liquid as shown in the schematic Fig. 3.1b then using the first law of thermodynamics [1–4] we may write

$$\mathrm{d}U = \mathrm{d}Q + \mathrm{d}W_{PV} + \mathrm{d}W_{NPV} = \mathrm{d}Q + \gamma \mathrm{d}A. \tag{3.2.1}$$

Assuming zero PV work. i.e. $\mathrm{d}W_{PV} = 0$ and considering non-PV work arising from an increase in surface area (A) against the surface tension (γ) forces where

$$\mathrm{d}W_{NPV} = \gamma \mathrm{d}A. \tag{3.2.2}$$

Considering an open system where particles and energy get exchanged the equation modifies to

$$\mathrm{d}U = \mathrm{d}Q + \mathrm{d}W = \mathrm{d}Q + \gamma \mathrm{d}A + \sum_i \mu_i \mathrm{d}n_i \tag{3.2.2}$$

where μ_i and n_i corresponds to the chemical potential and number of moles of the i-th species.

Considering a reversible system, $\mathrm{d}Q = T\mathrm{d}S$ hence, we get

$$\mathrm{d}U = T\mathrm{d}S + \gamma \mathrm{d}A + \sum_i \mu_i \mathrm{d}n_i. \tag{3.2.3}$$

Integrating Eq. (3.3.3) yields:

$$U = TS + \gamma A + \sum_i \mu_i n_i. \tag{3.2.4}$$

Again differentiating both sides of Eq. (3.3.4) we obtain

$$\mathrm{d}U = T\mathrm{d}S + S\mathrm{d}T + \gamma \mathrm{d}A + A\mathrm{d}\gamma + \sum_i \mu_i \mathrm{d}n_i + \sum_i n_i \mathrm{d}\mu_i.$$

Comparing Eqs. (3.3.3) and (3.3.4) we get

$$S\mathrm{d}T + A\mathrm{d}\gamma + \sum_i n_i \mathrm{d}\mu_i = 0 \tag{3.2.5}$$

which is the Gibbs–Duhem equation for an open system undergoing reversible changes [4].

At the Gibbs interface separating phases 1 (blue circles) and 2 (red circles) separated by an interface σ Fig. 3.1a by a dotted line we have $V^\sigma = 0$, hence the Gibbs–Duhem equation becomes

$$S^\sigma \mathrm{d}T + A^\sigma \mathrm{d}\gamma + \sum_i n_i^\sigma \mathrm{d}\mu_i = 0. \tag{3.2.6}$$

Under isothermal conditions $\mathrm{d}T = 0$, we get

$$A^\sigma \mathrm{d}\gamma + \sum_i n_i^\sigma \mathrm{d}\mu_i = 0 \tag{3.2.6}$$

Or

$$d\gamma = -\sum_i \frac{n_i^\sigma}{A^\sigma} d\mu_i = \sum_i \Gamma_i d\mu_i \tag{3.2.7}$$

where Γ_i is the adsorption coefficient for the i-th species at the σ interface.

For a solution consisting of a solvent (suffix 1) and a solute (suffix 2) we have $d\gamma = -\Gamma_1 d\mu_1 - \Gamma_2 d\mu_2$ for simplicity we can place the dividing line such that $\Gamma_1 = 0$ and in that case the equation reduces to $d\gamma = -\Gamma_2 d\mu_2$ and because $\mu = \mu_0 + \text{RT} \ln a_i$ where a_i is the chemical activity coefficient of the i-th species it can be conveniently replaced by its molar concentration c_i. For $i = 2$ we have, therefore,

$$\Gamma_2 = -\frac{c_2}{\text{RT}}\left(\frac{d\gamma}{dc_2}\right). \tag{3.2.8}$$

This equation forms the backbone of tensiometry that connects the dependence of surface tension with bulk concentration (c_2). Clearly, the adsorption coefficient or surface excess term (Γ) is dependent on the bulk concentration, and the negative sign indicates that as $\frac{d\gamma}{dc_2}$ decreases, Γ_2 increases [5–8].

3.3 Correlation Between Thermodynamic Parameters and Physical Variables

The state function enthalpy (H) is given as $H = U + PV$ which on differentiation yields

$$dH = dU + d(PV) = dQ + dW + d(PV) = dQ + VdP. \tag{3.3.1}$$

At constant pressure, we get

$$(dH)_P = dQ = C_P dT \tag{3.3.2}$$

where C_P is the specific heat of the material at constant pressure.

Again the expression for internal energy may be expressed as

$$dU = dQ + dW = dQ + PdV. \tag{3.3.3}$$

At constant volume, we get

$$(dU)_V = C_V dT \tag{3.3.4}$$

where

$$C_V \text{ is the specific heat of material at constant volume.} \tag{3.3.5}$$

The expression for Gibbs free energy for an open system is given by

$$G = U + PV - TS + \sum_i \mu_i n_i. \tag{3.3.6}$$

Differentiating and using the expression below.

As

$$\mathrm{d}U = \mathrm{d}Q + \mathrm{d}W = T\mathrm{d}S + \gamma \mathrm{d}A - P\mathrm{d}V + \sum_i \mu_i \mathrm{d}n_i \tag{3.3.7}$$

we get

$$\mathrm{d}G = \gamma \mathrm{d}A + \sum_i \mu_i \mathrm{d}n_i + V\mathrm{d}P - S\mathrm{d}T. \tag{3.3.8}$$

Assuming an isothermal and isobaric process and the number of particles in the system is conserved, we have

$$\mathrm{d}G = \gamma \mathrm{d}A \text{ or } \gamma = \left(\frac{\mathrm{d}G}{\mathrm{d}A}\right)_{T,p,n}. \tag{3.3.9}$$

Thus surface tension is a measure of the rate of change of Gibbs free energy per unit area.

Again

$$\begin{aligned} G &= H - ST = (U + PV) - ST \\ &= ST - PV + \gamma A + \sum_i \mu_i n_i - ST \\ &= \gamma A + \sum_i \mu_i n_i \end{aligned}. \tag{3.3.10}$$

Therefore

$$G = \gamma A + \sum_i \mu_i n_i \text{ and } H = ST + \gamma A + \sum_i \mu_i n_i. \tag{3.3.11}$$

According to the Gibbs theory of surfactant adsorption [3–7]

$$G^\sigma = H^\sigma - TS^\sigma = U^\sigma + PV^\sigma - TS^\sigma. \tag{3.3.11}$$

The differential form of the same is given by

$$dG^{\sigma} = dU^{\sigma} + PdV^{\sigma} + V^{\sigma}dP - TdS^{\sigma} - S^{\sigma}dT \tag{3.3.12}$$

and as

$$dU^{\sigma} = TdS^{\sigma} + \gamma dA - PdV^{\sigma} + \sum_i \mu_i dn_i \tag{3.3.13}$$

yields

$$dG^{\sigma} = +\gamma dA + \sum_i \mu_i dn_i + VdP - S^{\sigma}dT \tag{3.3.14}$$

which modifies to

$$dG^{\sigma} = +\gamma dA + \sum_i \mu_i dn_i. \tag{3.3.15}$$

At constant temperature, pressure, and at equilibrium

$$\gamma = \sum_i \mu_i \frac{dn_i}{dA} = \sum_i \mu_i \Gamma_i \tag{3.3.16}$$

where Γ_i is the adsorption coefficient of species i, and is identical to Eq. 3.3.8.

3.4 Dependence of Surface Tension on Temperature

The Gibbs free energy is given as

$$dG^{\sigma} = \gamma dA + \sum_i \mu_i dn_i + VdP - S^{\sigma}dT. \tag{3.4.1}$$

At equilibrium and constant pressure with a total number of particles conserved we have

$$dG^{\sigma} = +\gamma dA - S^{\sigma}dT. \tag{3.4.2}$$

Dividing by dT we get

$$\frac{dG^{\sigma}}{dT} = \gamma \frac{dA}{dT} - S^{\sigma} \tag{3.4.3}$$

and as A is a constant the expression modifies to $\frac{dG^{\sigma}}{dT} = -S^{\sigma}$. Dividing both sides by A yields $\frac{dG^{\sigma}/A}{dT} = -\frac{S^{\sigma}}{A}$ On integrating $dG^{\sigma} = +\gamma dA - S^{\sigma}dT$ we get

$$\frac{G^\sigma}{A} = \gamma - \frac{S^\sigma}{A}T. \tag{3.4.4}$$

3.5 Adsorption

The schematic Fig. 3.2 shows an isolated system consisting of n_i number of particles in phases 1 and 2 respectively, while n_σ the number of particles at the **σ** interface.

At equilibrium, the chemical potentials at the interface is μ_σ which is same as that in the bulk and is given by μ_1. Mathematically, these chemical potentials at the interface and in bulk of phase 1 are given by

$$\mu_1 = \mu_1^0 + \mathrm{RT}\ln c_1 \tag{3.5.1}$$

and

$$\mu_\sigma = \mu_\sigma^0 + \mathrm{RT}\ln c_\sigma. \tag{3.5.2}$$

Subtracting, the two equations we have

$$\mu_\sigma - \mu_1 = (\mu_\sigma^0 - \mu_1^0) + \mathrm{RT}\ln\frac{c_\sigma}{c_1}. \tag{3.5.3}$$

Because $\mu_\sigma = \mu_1$ at equilibrium we have

$$(\mu_\sigma^0 - \mu_1^0) + \mathrm{RT}\ln\frac{c_\sigma}{c_1} = 0. \tag{3.5.4}$$

Simplifying we have

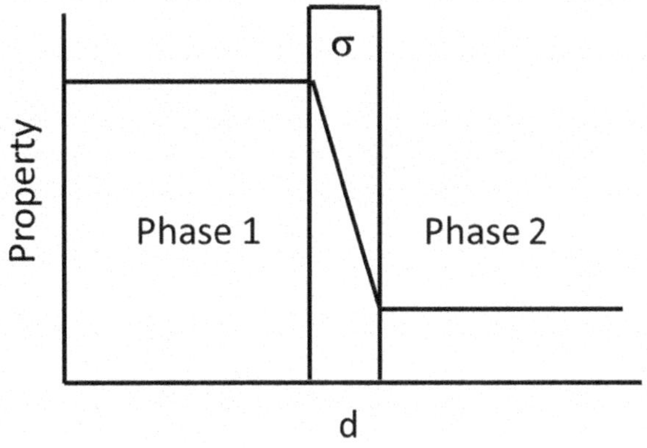

Fig. 3.2 Schematic of two phases (1 and 2) separated by an infinitesimally thin interface σ

$$c_{\sigma} = c_1 \exp\left(-\frac{\Delta G^0_{i\sigma}}{\mathrm{RT}}\right). \tag{3.5.5}$$

If γ_0 and γ are the surface tensions at the interface between fluids 1 and 2 then the surface excess is given by $\Gamma_i = \frac{n_i}{A}$ where n_i the number of molecules corresponding to species *I* and *A* is the area of the interface. The same may be alternatively expressed as

$$\Gamma_i = \frac{n_i}{A} = \frac{n_i\delta}{A\delta} = \frac{n_i}{V}\delta = c_i\delta \tag{3.5.6}$$

where δ is the thickness of the interface. Again surface excess may be related to surface tension by the relation.

Combining Eqs. (3.3.9) and (3.5.9) term we have

$$\frac{-c_i}{\mathrm{RT}}\left(\frac{\mathrm{d}\gamma}{\mathrm{d}c_i}\right) = c_i\delta \tag{3.5.7}$$

or

$$\gamma_0 - \gamma = -\delta\mathrm{RT}(c_i^0 - c_\sigma) = -\delta\mathrm{RT}c_\sigma \tag{3.5.8}$$

when $\gamma = \gamma_0$ and $c = c_0$ we get

$$\gamma_0 - \gamma = \delta\mathrm{RT}c_\sigma = c_i\delta\mathrm{RT}\exp\left(\frac{-\Delta G^0_{1\sigma}}{\mathrm{RT}}\right). \tag{3.5.9}$$

or

$$\Pi_\sigma = c_i\delta\mathrm{RT}\exp\left(\frac{-\Delta G^0_{1\sigma}}{\mathrm{RT}}\right). \tag{3.5.10}$$

This expression connects change is surface pressure to Gibbs free energy term *G* and the thickness of the interface [5, 7].

3.6 Micellization

A micelle may be considered to be an organized structure evolving from aggregation of surfactant molecules. The formation is such that hydrophobic-hydrophobic interaction between the chains causes clustering of the surfactant molecules to form a core that is hydrophobic in nature. While the hydrophobic chains cluster in this core, the hydrophilic end groups on the other hand are located on the outer surface of the sphere and in contact with the aqueous media. The chemical potential is defined as $\mu_i = \mu_i^0 + \mathrm{RT}\ln a_i$ where a_i is the activity and may be replaced by the concentration of species (c_i). Assuming low concentrations of the species, the concentration may be represented as a mole fraction X_i. Here μ_i^0 is the standard potential of the same species at temperature T °K.

Let us now suppose that N molecules aggregate together to form a micelle i.e. NM $\Leftrightarrow M_N$.

Then it follows that $N\mu(M)$ is the total chemical potential of N monomers and this should be equal to the total chemical potential of the micellar species. Therefore, at equilibrium, we have

$$N\mu_i = \mu^0_{\text{micellar}} + RT\ln X_{\text{monomer}} \tag{3.6.1}$$

which simplifies to

$$\mu^0_{\text{micellar}} - N\mu_i^0 = \mathrm{RT}(\ln X_{\text{micellar}} - N\ln X_{\text{monomer}}). \tag{3.6.2}$$

As X_{monomer} is negligibly small and hence $\ln X_{\text{monomer}}$ tends to zero. Therefore we have

$$\mu^0_{\text{micellar}} - n\mu_i^0 = \mathrm{RT}\ln X_{\text{micellar}}. \tag{3.6.3}$$

Putting $\mu^0_{micellar} - n\mu_i^0 = \Delta G^0_{micellar}$ we have

$$\Delta G^0_{\text{micellar}} = \mathrm{RT}\ln X_{\text{micellar}} = \mathrm{RT}\ln cmc. \tag{3.6.4}$$

It must be noted that the Gibbs free energy is connected to the enthalpy and entropy through the relation

$$\Delta G^0_{\text{micellar}} = \Delta H^0_{\text{micellar}} - T\Delta S^0_{\text{micellar}}. \tag{3.6.5}$$

Differentiating with respect to T we have

$$\frac{\mathrm{d}}{\mathrm{d}T}\left(\Delta G^0_{\text{micellar}}\right) = \frac{\mathrm{d}}{\mathrm{d}T}(\Delta H^0_{\text{micllar}}) - \Delta S^0_{\text{micellar}} - T\frac{\mathrm{d}}{\mathrm{d}T}(\Delta S^0_{\text{micellar}}) \tag{3.6.6}$$

when $\Delta H^0_{\text{micellar}}$ and $\Delta S^0_{\text{micellar}}$ are constants, we have

$$\frac{\mathrm{d}}{\mathrm{d}T}\left(\Delta G^0_{\text{micellar}}\right) = -\Delta S^0_{\text{micellar}}. \tag{3.6.7}$$

Again starting from the equation

$$\Delta G^0_{\text{micellar}} = \Delta H^0_{\text{micellar}} - T\Delta S^0_{\text{micellar}}. \tag{3.6.8}$$

We get

$$\frac{\mathrm{d}}{\mathrm{d}T}\left(\frac{\Delta G^0_{\text{micellar}}}{T}\right) = \frac{\mathrm{d}}{\mathrm{d}T}\left(\frac{\Delta H^0_{\text{micellar}}}{T}\right) - \frac{\mathrm{d}}{\mathrm{d}T}\left(\Delta S^0_{\text{micellar}}\right). \quad (3.6.9)$$

The third term reduces to zero as $\Delta S^0_{\text{micellar}}$ is a constant. Therefore, the expression reduces to

$$\frac{\mathrm{d}}{\mathrm{d}T}\left(\frac{\Delta G^0_{\text{micellar}}}{T}\right) = -\frac{\Delta H^0_{\text{micellar}}}{T^2} \text{ or } \Delta H^0_{\text{micellar}} = -T^2\frac{\mathrm{d}}{\mathrm{d}T}\left(\frac{\Delta G^0_{\text{micellar}}}{T}\right). \quad (3.6.10)$$

The importance of this equation stems from the fact that the enthalpy of micellization can be determined calorimetrically allowing the evaluation of Gibbs free energy. Experimental measurements show that $\Delta H^0_{\text{micellar}}$ (kJ/mol) can be positive or negative but because it follows that $\Delta G^0_{\text{micellar}} < 0$ implies spontaneous formation of micelles in the solution. The positive value of $\Delta S^0_{\text{micellar}}$ therefore suggests that the process of micellization is entropy-driven [5, 9].

3.7 Role of Curvature

If we consider a spherical air bubble of radius R formed in bulk liquid (Fig. 3.3) and the air pressure inside is $P_0 + \Delta P$ while the pressure outside is P_0 then excess pressure inside bubble is ΔP and hence P–V work needed to increase its radius from R to $R + \mathrm{d}R$ is

$$\mathrm{d}W_{PV} = \Delta P\mathrm{d}V = \frac{4}{3}\pi\Delta P\mathrm{d}(R^3) = 4\pi R^2\Delta P\mathrm{d}R. \quad (3.7.1)$$

Now the Helmholtz Free Energy (F) is given by

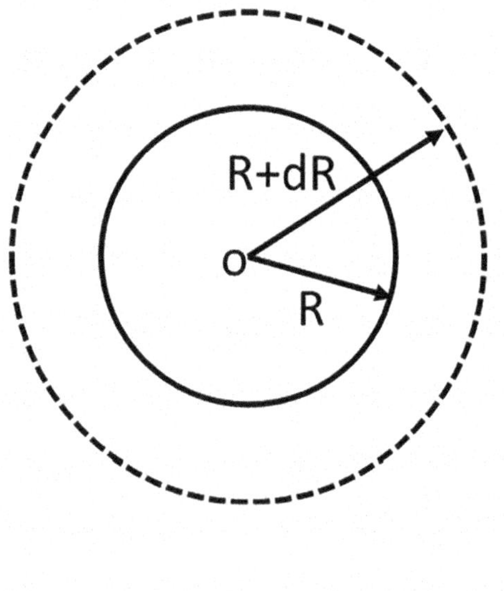

Fig. 3.3 Schematic of a spherical bubble of radius R

$$dF = d(U - ST) = -PdV + \sum_i \mu_i dn_i + \gamma dA - SdT. \quad (3.7.2)$$

Considering equilibrium ($dF = 0$) no transfer of particles ($dn_i = 0$) and isothermal conditions ($dt = 0$).we get

$$\Delta P dV = \gamma dA \quad (3.7.3)$$

or

$$\Delta P = \frac{2\gamma}{R}. \quad (3.7.4)$$

3.8 Thermodynamic Derivation of Young's Equation and Zisman Curves

The interfacial tensions acting at the liquid–vapor interface is γ_{LV}, at the solid–liquid and solid–vapor interfaces are γ_{SL} and γ_{SV} respectively. Therefore, the work done in shifting the triple line through an elementary distance dx along PP′ is given by

$$dW = F.dx = (\gamma_{SL} + \gamma_{LV}\cos\theta - \gamma_{SV})dx \quad (3.8.1)$$

and because at equilibrium $dW = 0$we get

$$\cos\theta_Y = \frac{\gamma_{SA} - \gamma_{SL}}{\gamma_{LA}} \quad (3.8.2)$$

which is Young's equation and $\theta = \theta_Y$ at equilibrium. As is evident from Young's equation $\cos\theta_Y = \frac{\gamma_{SA} - \gamma_{SL}}{\gamma_{LA}}$ the terms in the numerator cannot be evaluated directly although their difference can be evaluated with accuracy. Zisman noticed that when the contact angle of various solvent drops on a specific substrate was plotted against its surface tension, it yielded a straight line. Extrapolating this line intersects the line at P (Fig. 3.4b) corresponding to $\cos\theta = 1$, a condition for complete wetting of the substrate. This value of $\gamma = \gamma_C$ at which $\cos\theta = 1$ suggests that any solvent with surface tension γ_C will always wet that specific surface. Interestingly however the value of γ_C is independent of the solvent which is very strange and has invited much criticism. Nevertheless, the above holds good for low-energy surfaces like paraffin, polyethylene, etc.

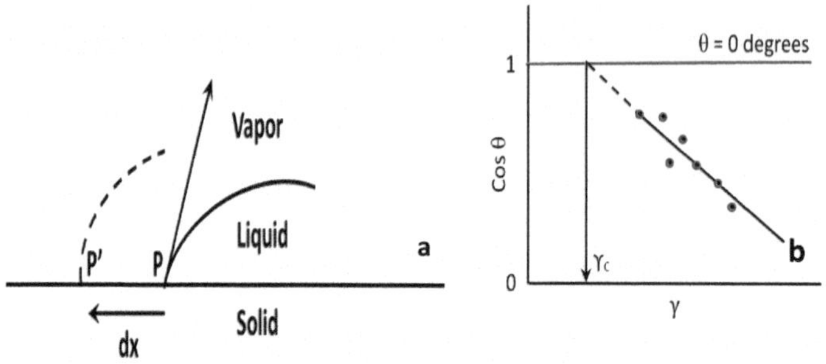

Fig. 3.4 **a** Schematic of a liquid drop at equilibrium on a solid substrate. *P* corresponds to the triple point, **b** Zisman curve

3.9 Wetting of Solid Surfaces

Wetting of substrates is of great significance as it determines whether or not a given molecule will wet and to a surface which is crucial in the design of a wide range of products from cosmetics to pharmaceutical drugs [10–12]. Consider a smooth inert substrate of surface energy γ_{SV} and a droplet of a liquid having surface energy γ_{LV} being deposited on to the substrate wetting it and the contact angle is θ as shown in Fig. 3.5.

Assuming the film of liquid to be thin and uniform, the net energy of the substrate is

$$W_2 = \gamma_{SV}(A_{Tot} - A_{SL}) + \gamma_{SL}A_{SL} + \gamma_{LV}A_{SL} \tag{3.9.1}$$

where A_{SL} is the wetted area and $W_1 = \gamma_{SV}A_{Tot}$ change in work done to form the film on the substrate.

$$\Delta W_{adhesion} = W_2 - W_1 = (\gamma_{SV} - \gamma_{SL} + \gamma_{LV})A_{SL} = \gamma_{LV}(1 + \cos\theta)A_{SL} \tag{3.9.2}$$

thus work of adhesion per unit area is

$$\Delta W_{adhesion} = W_2 - W_1 = \gamma_{LV}(1 + \cos\theta). \tag{3.9.3}$$

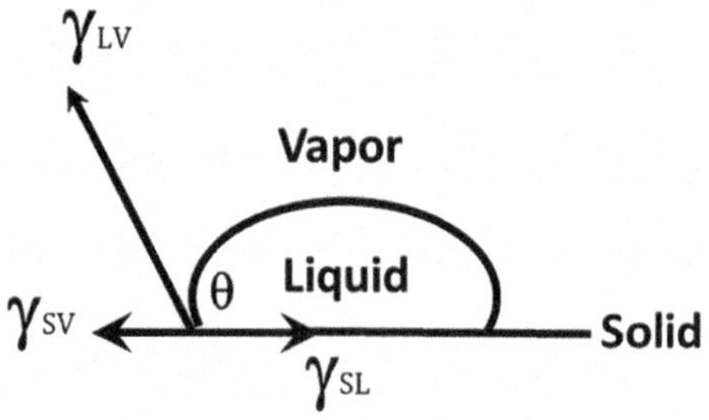

Fig. 3.5 Schematic of a liquid droplet on a smooth solid substrate at equilibrium. γ_{LV}, γ_{SV} and γ_{SL} corresponds to the surface tensions at the liquid–vapor, solid–liquid, and solid–liquid interfaces

For complete wetting,

$$\theta = 0^\circ \Delta W_{\text{adhesion}} = 2\gamma_{\text{LV}}. \tag{3.9.4}$$

For complete de-wetting or superhydrophobicity

$$\theta = 180^\circ, \Delta W_{\text{adhesion}} = 0. \tag{3.9.5}$$

And for partial wetting

$$\theta = 90^\circ W_{\text{adhesion}} = \gamma_{\text{LV}}. \tag{3.9.6}$$

3.10 Adhesion and Cohesion Coefficients

Let us consider a liquid drop in the shape of a spherical cap to be at equilibrium on a solid substrate. The base area of the spherical cap is A_{SL}, surface are of the cap is A_{LV}, and uncovered area on the flat plate is $A_{\text{SV}} = A_{\text{Total}} - A_{\text{SL}}$ where A_{Total} is the area of the plate shown in Fig. 3.6.

If we consider a liquid drop at equilibrium on a solid substrate as shown in Fig. 3.7, then at equilibrium, the total energy of the system E is given by:

$$\begin{aligned} E_1 &= \gamma_{\text{SL}} A_{\text{SL}} + \gamma_{\text{SV}} A_{\text{SV}} + \gamma_{\text{LV}} A_{\text{LV}} \\ &= \gamma_{\text{SL}} A_{\text{SL}} + \gamma_{\text{SV}} (A_{\text{Total}} - A_{\text{SL}}) + \gamma_{\text{LV}} A_{\text{LV}}. \end{aligned} \tag{3.10.1}$$

When the liquid spreads then

$$\begin{aligned} E_2 &= \gamma_{\text{SL}} A'_{\text{SL}} + \gamma_{\text{SV}} A'_{\text{SV}} + \gamma_{\text{LV}} A'_{\text{LV}} \\ &= \gamma_{\text{SL}} A'_{\text{SL}} + \gamma_{\text{SV}} (A_{\text{Total}} - A'_{\text{SL}}) + \gamma_{\text{LV}} A'_{\text{LV}}. \end{aligned} \tag{3.10.2}$$

Thus work done during the spreading of the liquid is

$$\begin{aligned} E_2 - E_1 = \gamma_{\text{SL}} &(A'_{\text{SL}} - A_{\text{SL}}) \\ &- \gamma_{\text{SV}} (A'_{\text{SL}} - A_{\text{SL}}) + \gamma_{\text{LV}} (A'_{\text{LV}} - A_{\text{LV}}) \end{aligned} \tag{3.10.3}$$

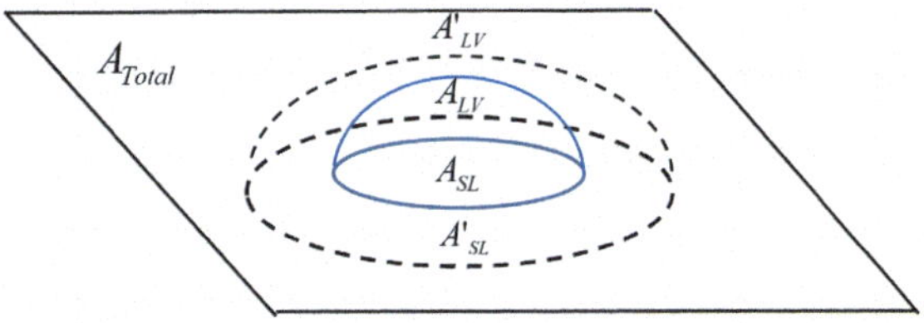

Fig. 3.6 Schematic of a spherical liquid cap on a substrate

$$\Delta E = (\gamma_{SL} - \gamma_{SV})(A'_{SL} - A_{SL}) + \gamma_{LV}(A'_{LV} - A_{LV})$$
$$= -\gamma_{LV}\cos\theta(\Delta A_{SL}) + \gamma_{LV}(\Delta A_{LV}) \quad (3.10.4)$$

$$\Delta E = -\gamma_{LV}\cos\theta(\Delta A_{SL}) + \gamma_{LV}(\Delta A_{LV})$$
$$= -\gamma_{LV}(1 + \cos\theta)\Delta A_{SL}. \quad (3.10.5)$$

Therefore, work of adhesion

$$\Delta W_{Adhesion} = \frac{\Delta E}{\Delta A_{SL}} = -\gamma_{LV}(1 + \cos\theta)\Delta A_{SL} \quad (3.10.6)$$

as

$$\Delta A_{SL} + \Delta A_{LV} = 0 \quad (3.10.7)$$

that can be readily shown given the drop to be a spherical cap and change of contact angle $\Delta\theta$ is small.

Volume of the cap is given by

$$V_{Cap} = \frac{2\pi}{3}r^3(1 - \sin\theta). \quad (3.10.8)$$

Surface area of the cap

$$A_{Cap} = 2\pi r^2(1 - \sin\theta). \quad (3.10.9)$$

Base area of cap is

$$A_{Base} = 2\pi r^2\cos\theta. \quad (3.10.10)$$

The negative sign in the expression ΔE_{12} indicates that ΔE_{12} decreases as ΔA_{SL} increases.

At $\theta = 0$, it follows that

$$W_{Adhesion} = -2\gamma_{LV}. \quad (3.10.11)$$

At $\theta = 90$, it follows that

$$W_{Adhesion} = -\gamma_{LV}. \quad (3.10.12)$$

At $\theta = 180$, it follows that

$$W_{Adhesion} = 0. \quad (3.10.13)$$

For cohesion between two liquid drops, we get $W_{\mathrm{LL}} = \gamma_{\mathrm{LV}} + \gamma_{\mathrm{LV}} = 2\gamma_{\mathrm{LV}}$ and from the above two relations we get

$$\cos\theta = \left(2\frac{W_{\mathrm{SL}}}{W_{\mathrm{LL}}} - 1\right). \quad (3.10.14)$$

3.11 Spreading Coefficient (S)

The spreading coefficient (S) is given by the difference in energy of the substrate when dry and when wet and is mathematically expressed as $S = (\pi_{S0}) - (\pi_{\mathrm{SL}} + \pi_{\mathrm{LV}})$ where π_{S0} corresponds to the surface energy of the solid substrate when dry i.e., space above the substrate is vacuum as indicated by the suffix 0. Wet corresponds to the space above the substrate saturated with vapor as indicated by the suffix v. π_{SL} and π_{LV} corresponds to the surface energies of the solid–liquid (SL) and the liquid–vapor (LV) interface, respectively. S will be positive when $\pi_{S0} > (\pi_{\mathrm{SL}} + \pi_{\mathrm{LV}})$. Because the surface energies are numerically equal to the surface tensions we have

$$\pi_{S0} = \gamma_{S0};\ \pi_{\mathrm{LV}} = \gamma_{\mathrm{LV}} \text{ and } \pi_{\mathrm{SL}} = \gamma_{\mathrm{SL}}. \quad (3.11.1)$$

Therefore, the equation modifies to

$$\begin{aligned} S &= (\gamma_{\mathrm{SV}}) - (\gamma_{\mathrm{SL}} + \gamma_{\mathrm{LV}}) \\ &= (\gamma_{\mathrm{SV}} - \gamma_{\mathrm{SL}}) - \gamma_{\mathrm{LV}} = \gamma_{\mathrm{LV}}(\cos\theta - 1). \end{aligned} \quad (3.11.2)$$

At $\theta = 0°$, i.e. when complete wetting occurs we have

$$S = 0. \quad (3.11.3)$$

At $\theta = 90°$, i.e. when partial wetting occurs we have

$$S = -\gamma_{Lv}. \quad (3.11.4)$$

At $\theta = 180°$, i.e. when complete de-wetting occurs we have

$$S = -2\gamma_{LV}. \quad (3.11.5)$$

Thus the highest possible value of the spreading coefficient is zero which corresponds to complete wetting.

3.12 Dependence of Spreading on Polarizability of a Solid

If we consider two infinite solids brought in close contact with each other. The work of adhesion then is

$$W = 2\gamma_{SO}. \tag{3.12.1}$$

And if Van der Waals (V_{SS}) is the only contributor then

$$W = 2\gamma_{SO} - V_{SS} = 0 \text{ or } 2\gamma_{SO} = V_{SS}. \tag{3.12.2}$$

When the solid and a liquid layer are brought together we have

$$\gamma_{\mathrm{SL}} = \gamma_{\mathrm{LO}} + \gamma_{\mathrm{SO}} - V_{\mathrm{SL}} \tag{3.12.3}$$

and when the two liquid interfaces are brought together we have

$$W = 2\gamma_{LO} - V_{LL} = 0 \text{ or } 2\gamma_{LO} = V_{LL}. \tag{3.12.4}$$

Plugging these values in Eq. (3.11.4), we get

$$S = \gamma_{SA} - (\gamma_{LS} + \gamma_{LA}) = V_{SL} - V_{LL} = k\alpha_L(\alpha_S - \alpha_L) \tag{3.12.5}$$

as $V_{SL} = k\alpha_S\alpha_L$ and $V_{LL} = k\alpha_L^2$ where α_S and α_L are the polarizabilities in the solid and liquid.

Clearly, the above expression shows that spreading a liquid on a solid surface depends on the difference between the polarizability of the solid and the liquid. Hence, when $\alpha_S > \alpha_L$ then S > 0 and wetting is total that leads to the thumb rule that when the liquid is less polarizable than the solid then the liquid spreads totally over the solid which explains why helium despite its negligible polarizability spreads over almost all solids [12].

Low-energy solids namely molecular crystals and plastics are hardly wetted by water. This is because the chemical binding energies of these solids are of the order of k_BT (4.2×10^{-21} J) while high-energy solids have chemical binding energies of the order of 1 eV (1.6×10^{-19} J).

In high energy solids $\gamma_{SA} > \gamma_{LS}$ then $\cos\theta_Y > 0$ implying $0 < \theta_Y < 90$ that explains why high energy solids are well wetted by liquids, e.g., metals, ionic solids.

3.13 Relation Between Adsorption and Contact Angle

Let us consider a surface of area (A) in vacuum and surface energy (γ_0) be exposed to a vapor of some material whereupon adsorption of some vapor molecules occur on the surface of the substrate and the net free energy of the surface is given by the relation:

$$F = \gamma_0 - \sum_i \mu_i \Gamma_i. \tag{3.13.1}$$

Differentiating we have

$$\mathrm{d}F = \sum_i \mu_i \mathrm{d}\Gamma_i + \Gamma_i \mathrm{d}\mu_i. \tag{3.13.2}$$

Since the chemical potential is a constant we have $\mathrm{d}\mu_i = 0$. Then,

$$\mathrm{d}F = -\sum_i \mu_i \mathrm{d}\Gamma_i. \tag{3.13.3}$$

Therefore energy change in Free energy for adsorption of species from the vapor phase on to the solid is given by

$$F_1 = \gamma_{SO} - (\gamma_{\mathrm{SV}} - \mu_{\mathrm{SV}}\Gamma_{\mathrm{SV}}). \tag{3.13.4}$$

Similarly, for the solid–liquid interface we have

$$F_2 = \gamma_{SO} - (\gamma_{\mathrm{SL}} - \mu_{\mathrm{SL}}\Gamma_{\mathrm{SL}}) \tag{3.13.5}$$

and for liquid–gas interface, we have

$$F_3 = \gamma_{LO} - (\gamma_{\mathrm{LV}} - \mu_{\mathrm{LV}}\Gamma_{\mathrm{LV}}) \tag{3.13.6}$$

$$\begin{aligned} F_1 - F_2 &= \gamma_{\mathrm{SL}} - \gamma_{\mathrm{SV}} + \mu_{\mathrm{SV}}\Gamma_{\mathrm{SV}} - \mu_{\mathrm{SL}}\Gamma_{\mathrm{SL}} \\ &= -\gamma_{\mathrm{LV}}\cos\theta + \mu_{\mathrm{SV}}\Gamma_{\mathrm{SV}} - \mu_{\mathrm{SL}}\Gamma_{\mathrm{SL}}. \end{aligned} \tag{3.13.7}$$

At equilibrium, $F_1 - F_2 = 0$. Therefore,

$$\mu_{\mathrm{SV}}\Gamma_{\mathrm{SV}} - \mu_{\mathrm{SL}}\Gamma_{\mathrm{SL}} = \gamma_{\mathrm{LV}}\cos\theta. \tag{3.13.8}$$

Because the liquid and liquid–vapor is in equilibrium $\mu_{\mathrm{SV}} = \mu_{\mathrm{SL}} = \mu_i$ being the same molecule, it has the same chemical potential. Therefore,

$$\frac{\mathrm{d}(\gamma_{LV}\cos\theta)}{\mathrm{d}\mu_i} = \Gamma_{SV} - \Gamma_{SL}. \tag{3.13.9}$$

3.14 Leidenfrost Effect

Leidenfrost had observed that when drops of water come in contact with a hot surface (220 °C and above) the drops roll into spheres that bump off from the surface. He concluded that the liquid surface that came into contact with the hot surface rapidly evaporated to form a thin film that is about 0.1–0.2 mm thick causing the drops to bump off the drop from the hot surface (Fig. 3.7).

If the drop has radius (R) and the radius of the area of contact with the surface is r, then the pressure difference arising from the curvature of the drop is given by

$$\Delta P = \frac{2\gamma}{R}. \tag{3.14.1}$$

And the force on the drop is given by

$$F = \pi r^2 \Delta P = 2\pi r^2 \frac{\gamma}{R}. \tag{3.14.2}$$

That balances the weight (mg) of the drop and hence

$$2\pi r^2 \frac{\gamma}{R} = mg \tag{3.14.3}$$

which on simplification yields

$$r^2\gamma = \frac{2}{3} R^4 \rho_l g \tag{3.14.4}$$

that yields

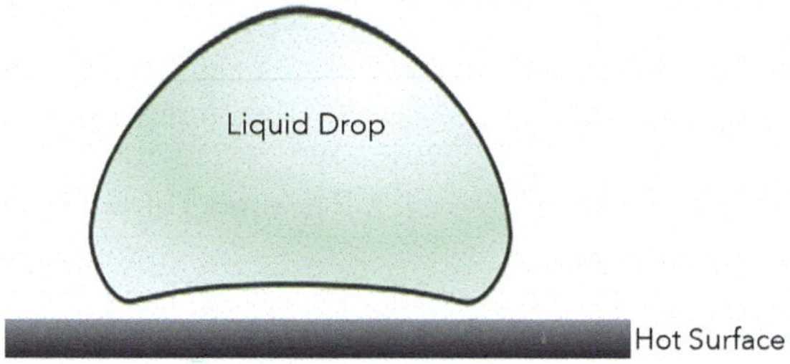

Fig. 3.7 Schematic drop of a liquid placed on a hot plate heated to above 220 °C

$$\frac{r^2}{R^2} = \frac{2}{3\gamma} R^2 \rho g. \tag{3.14.5}$$

Because $\sin\theta = \frac{r}{R}$, we get

$$\sin\theta = R\left(\frac{2\rho g}{3\gamma_{\mathrm{LA}}}\right)^{\frac{1}{2}} \tag{3.14.6}$$

where ρ is the density of the liquid and g is the acceleration due to gravity.

3.15 Contact Angle on Rough Substrates

Liquid surfaces are different from solid surfaces as the Brownian motion of the molecules in a liquid repairs any local defect created and hence at equilibrium, liquid surfaces are always molecularly smooth. This however is not the case for solids. Imperfections like holes, crystal defects, dislocations, and ridges produced by overlapping crystal planes make the surface of solids microscopically rough. Because of their jagged surface, the tangent made with the LV interface at the triple point varies from point to point on the substrate and hence the observed contact angle (θ^*) depends on its location on the substrate. It must be pointed out that the observed contact angle (θ^*) is different from Young's contact angle (θ_Y) which corresponds to the contact angle on a smooth surface Fig. 3.8a, b.

Let us consider an elementary surface area $\mathrm{d}A_{\mathrm{LA}}$ that subtends a projected area $\mathrm{d}A_F$ on a flat smooth surface.

It follows from simple geometry that

$$\cos\theta^* = \frac{\mathrm{d}A_{\mathrm{LA}}}{\mathrm{d}A_F}. \tag{3.15.1}$$

The same may be rewritten as

$$\cos\theta^* = \frac{\mathrm{d}A_{\mathrm{LA}}}{\mathrm{d}A_F} = \left(\frac{\mathrm{d}A_{\mathrm{SL}}}{\mathrm{d}A_F}\right)\left(\frac{\mathrm{d}A_{\mathrm{LA}}}{\mathrm{d}A_{\mathrm{SL}}}\right) = r\cos\theta \tag{3.15.2}$$

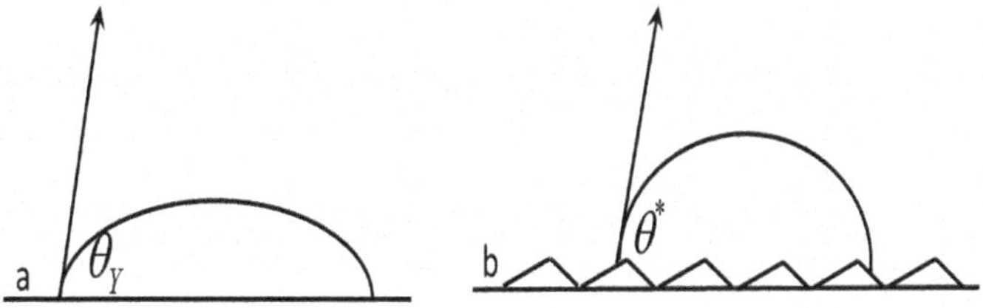

Fig. 3.8 Schematic of a liquid drop on **a** a smooth solid substrate and **b** a corrugated solid surface

where 'r' is the roughness factor and is given by $r = \frac{dA_{SL}}{dA_F}$ that represents the ratio of the surface area of the solid–liquid interface to its projection on a perfectly smooth flat surface (F) while $\cos\theta$ represents the ratio of surface area at the liquid–air interface to its projection on the solid–liquid interface and is always greater than unity. This relationship connecting the observed contact angle on a rough surface (θ^*) and the contact angle between the same pair of solid and liquid when the solid surface is smooth (θ_Y) is referred to as Wenzel's relation [5, 7, 10, 12] that may be also derived from thermodynamic considerations as discussed in the next section.

3.16 Thermodynamic Derivation of Wenzel's Relation

As mentioned earlier, Young's equation holds good for a smooth interface and is given as $\gamma_{SA} - \gamma_{SL} = \gamma_{LA}\cos\theta_Y$. This expression does not hold good for an inhomogeneous surface containing both physical defects and a heterogeneous chemical composition. Assuming the surface energy of a rough surface to be $r(\gamma_{SA} - \gamma_{SL})$ where r is the roughness factor, we may write and expression for the total differential work done in forming the drop having an apparent contact angle θ^* as

$$dE_W = \gamma_{LA}\cos\theta^* l dx + rl(\gamma_{SA} - \gamma_{SL})dx \tag{3.16.1}$$

where l is the length and dx is the width of an elementary strip on an interface.

Putting $\frac{dE_W}{dx}$ equal to zero at equilibrium,we get

$$\gamma_{LA}\cos\theta^* + r(\gamma_{SA} - \gamma_{SL}) = 0. \tag{3.16.2}$$

Plugging in Young's equation we have

$$\cos\theta^* = r\left(\frac{\gamma_{SL} - \gamma_{SA}}{\gamma_{LA}}\right) = r\cos\theta_Y \tag{3.16.3}$$

which is Wenzel's equation. This relation is of importance as it connects the apparent contact angle (θ^*) to the roughness (r) of the solid–liquid interface. For smooth surfaces, $r = 1$, $\theta^* = \theta_Y$.

For rough surfaces, $r > 1$ and for a hydrophobic liquid $\theta_Y > 90°$, $\theta* > \theta_Y$ that is a hydrophobic liquid tends to become more hydrophobic which increases with increasing roughness.

Similarly, for a hydrophilic liquid $\theta_Y < 90°$, $\theta* < \theta_Y$ that is a hydrophilic liquid tends to become more hydrophilic and hydrophilicity increases with roughness [13].

Roughness of material surfaces may not only originate from physical defects such as holes, crystal lattice defects, and surface textures but also from the existence of domains having different chemical compositions. Such domains may originate from effects like chemical leaching, oxidation, or reduction of chemical species constituting the surface, and adsorption of other molecules on its surface. Considering a heterogeneous surface having at least two components contributing to the roughness and hence two different contact angles, we have Cassie's equation as

$$\cos\theta = f_1 \cos\theta_1 + f_2 \cos\theta_2 \tag{3.16.4}$$

where θ_1 and θ_2 are the contact angles corresponding to area fractions f_1 and f_2. Here $f_1 + f_2 = 1$. Therefore, for a multicomponent system, Cassie's equation may be modified as $\cos\theta^* = \sum_{i=1}^{n} f_i \cos\theta_i$ where f_i corresponds to the fractional area corresponding to the i-th material and θ_i is the corresponding contact angle.

3.17 Limitations of the Wenzel and Cassie-Baxter Equations

The equation $\cos\theta = f_1 \cos\theta_1 + f_2 \cos\theta_2$ assumes that the heterogeneous surface consists of distinct domains that are well separated and their free surface energy may be averaged. It has been pointed out that when the domains are of atomic dimensions then instead of their free energies the dipole moment of a macromolecule needs to be determined. Then the equation modifies to $(1 + \cos\theta)^2 = f_1(1 + \cos\theta_1)^2 + f_2(1 + \cos\theta_2)^2$ which stands verified for polymers with different functional groups [12, 13]. Moreover, it now stands established that the contact angle varies between θ_{Adv} and θ_{Rec}. The contact angle tends to advancing contact angle when liquid is added to the droplet and toward a receding angle when the liquid is withdrawn from the droplet. The difference between the advancing and receding contact angles is called the contact angle hysteresis that reflects the fundamental asymmetry of wetting and de-wetting processes and their irreversibility. Since atomically smooth surfaces also have some heterogeneity and roughness contact angle hysteresis always exist although it may be made very small $\delta\theta < 1°$ by carefully controlling the surface smoothness of surfaces. The contact angle hysteresis is also a measure of energy dissipation during flow of a droplet on a solid surface.

References

1. Kondepudi D (2008) Introduction to modern thermodynamics. Wiley, N.Y.
2. Luscombe JH (2018) Thermodynamics. CRC Press, Boca-Raton
3. Honig JM (2021) Thermodynamics: principles characterizing physical and chemical processes, 5th edn. Academic Press, San Deigo
4. Sherwood D, Dalby P (2018) Thermodynamics for chemists and biochemists. Oxford University Press, Oxford
5. Erbil HY (2006) Surface chemistry of solid and liquid interface. Wiley, N.Y.
6. Saggion A, Faraldo R, Pierno M (2019) Thermodynamics: fundamental principles and applications. Springer, N.Y.
7. Butt JH, Graf K, Kappl M (2003) Physics and chemistry of interfaces. Wiley-VCH, Berlin

8. Meier GH (2014) Thermodynamics of surfaces and interfaces: concepts in Inorganic materials. Cambridge University Press, Cambridge
9. Moroi Y (2005) Micelle: theoretical and applied aspects. Springer, New Delhi
10. Bormashenko EY (2017) Physics of wetting: phenomena and applications of fluids on surfaces. De Gruyter, Germany
11. Daoud M, Willams CE (eds) (1999) Soft matter physics. Springer-Verlag, Germany
12. de Gennes PG (1985) Rev Mod Phys 57:827–861
13. Nosonovsky M, Bhusan B (2008) Multiscale dissipative mechanisms and hierarchical surfaces. Springer, Berlin

Chapter 4
Experimental Methods

4.1 Introduction

In recent years, research in "Soft Matter" also known as "Complex Fluids" has gained enormous momentum due to their diverse applications in solid-state physics, chemistry, biology, and material sciences. These materials have properties intermediate to that of rigid solids and liquids. Liquid crystalline materials are excellent examples of soft materials that find industrial applications in liquid crystal displays (LCD) used in mobile phones, computers, and television screens that constitute a multibillion-dollar industry. Synthetic and biopolymers are also important members of the soft matter family that find numerous applications in food cosmetics and medical drug delivery systems. Despite excellent theoretical progress, given the complexity of real systems and overlap of diverse fields like physics, chemistry, biology, and material sciences, much of the effort continues to be experimental and requires the application of different experimental techniques hence the need for this chapter especially for a newcomer to this area of research. This chapter is broadly divided into four Sects. 4.2.1–4.2.4.

In Sects. 4.2.1–4.2.4, experimental techniques for examining solid–liquid, liquid–liquid and liquid–air interfacial properties like contact angle (θ), interfacial tension (γ) and surface pressure (π) versus area per molecule isotherms are provided. Also presented are the formulas for calculating adsorption and diffusion coefficient from interfacial tension measurements.

In Sects. 4.3.1–4.3.4, methods for measurement of electrical conductivity (σ), zeta potential (ζ), and viscosity (η) of colloidal systems are discussed.

In Sects. 4.4.1 and 4.4.2, thermal properties of materials using Differential Scanning Calorimetry (DSC) and Thermal Gravimetric Analyzer (TGA) are discussed. These measurements provide much information on the stability and phase transitions occurring in these systems.

In Sects. 4.5.1–4.5.4, various imaging techniques used to explore the surface and bulk properties of materials are discussed in some detail. Bright-field and polarized

A. K. Dutta, *Soft Matter*, https://doi.org/10.1007/978-981-96-0624-5_4

optical microscopy is often used to understand better the organization of molecules both in bulk as well as on the surface of materials. Scanning Electron Microscopy (SEM) is used to examine the surface texture of materials, polarized fluorescence, and Brewster angle microscopic techniques are often employed to examine the organization and alignment of molecules at interfaces. Transmission Electron Microscopy (TEM) provides info on the distribution of different species in bulk of the material while Atomic Force Microscopy (AFM) provides info on the orientation and organization as well as distribution of species on surfaces or interfaces. Such information is useful in understanding the structure–property relationship in various complex systems.

4.2 Interfacial Properties

4.2.1 Contact Angle Measurements

When a drop of a low-volatile liquid is placed on the surface of a flat substrate, it is observed that the drop shape depends on the liquid and the substrate material. The angle (θ_Y) made by the tangent drawn at the triple point of contact *P* where the vapor, liquid and solid phases meet and is at thermodynamic equilibrium. It must be noted that the contact angle θ_Y as shown in Fig. 4.1a is specific for a given pair of liquid and substrate.

Considering equilibrium and balancing the interfacial forces at the triple point *P*, yields the Dupre–Young equation

$$\gamma_{\mathrm{lv}} \cos \theta_Y = \gamma_{\mathrm{sv}} - \gamma_{\mathrm{sl}} \tag{4.2.1.1}$$

that allows the evaluation of $\gamma_{\mathrm{sv}} - \gamma_{\mathrm{sl}}$ where γ_{sv}, γ_{sl} and γ_{lv} corresponds solid–vapor, solid–liquid and liquid–vapor interfacial tensions where subscripts '*s*', '*l*' and '*v*' stand for solid, liquid, and saturated vapor above the liquid droplet and solid substrate.

While γ_{lv} may be readily measured γ_{sv} and γ_{sl} cannot be evaluated directly. Contact angle measurements are readily made by a simple telescope with a camera interfaced

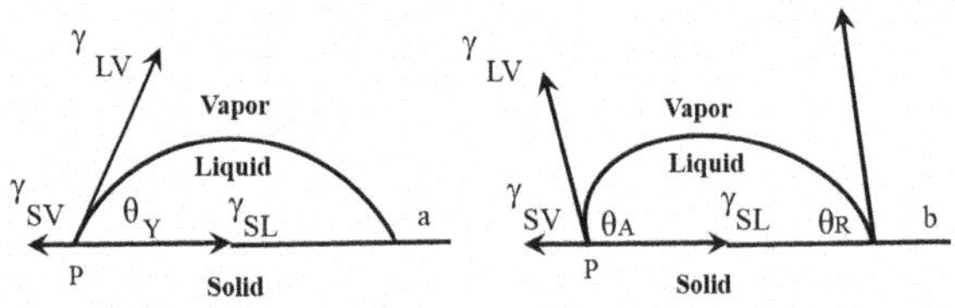

Fig. 4.1 **a** Schematic of a liquid droplet on a solid substrate where in (**a**) θ_Y corresponds to the Young's equilibrium angle, **b** θ_A and θ_R are the advancing and retreating angles. *P* is the triple point

through a computer that acquires and analyses the data. A small droplet of volume 10 μl is placed on a clean substrate of interest using a micro-syringe and an image of the droplet is acquired by the camera and analyzed. It must be noted that the contact angle θ_Y is a constant for a given liquid and substrate but depends on various factors like purity and volume of the liquid drop, cleanliness and roughness of the substrate. Detailed studies show that contact angles are functions of time. While the larger angle is called the advancing angle θ_A the other is called the retreating angle θ_R as shown in Fig. 4.1b. For measurements at any point in time, the average of the advancing and retreating angles are usually considered [1–3]. Contact angle measurements are important as it helps identify whether or not a liquid will spread on a substrate. For complete wetting, $\theta_Y = 0°$ that occurs when $\gamma_{\text{lv}} = \gamma_{\text{sv}} - \gamma_{\text{sl}}$, and for complete de-wetting, $\theta_Y = 90°$ that occurs when $\gamma_{\text{sv}} = \gamma_{\text{sl}}$. However, when $\theta_Y = 180°$ $\cos\theta_Y = -1$, $\gamma_{\text{lv}} = \gamma_{\text{sl}} - \gamma_{\text{sv}}$ the substrate is super hydrophobic surface.

4.2.2 Steady State Surface Tension at Liquid–Vapor Interfaces

Molecules located deep inside the volume of a liquid as shown in Fig. 4.2a experience net zero force of attraction or repulsion due to equal but oppositely directed components arising out of molecular interactions with the molecules surrounding it. However, molecules located at the vapor–liquid interface experiences unequal forces and hence a resultant force at the interface known as interfacial tension exists that make the liquid surface behave like a tight membrane stretched across the vapor–liquid interface. A simple and reliable means of measuring interfacial tension at the liquid–vapor interface is the Wilhelmy plate method where a clean and flat pure platinum plate is attached to the arm of a sensitive microbalance and its weight is noted or tared. The platinum plate is then partially dipped in the liquid as shown in Fig. 4.2b and the change in weight ΔW is recorded and the surface tension is calculated from the equation

$$\gamma = \frac{\Delta W}{2(b + t)\cos\theta_Y} \tag{4.2.2.1}$$

where θ_Y is the contact angle made by the platinum plate with the liquid meniscus, b and t are the width and thickness of the platinum plate respectively. Note that this expression is independent of the length (l) of the platinum plate. Accuracy of the above method is of the order of 0.01 mN/m and depends on the cleanliness of the platinum plate. To clean the Wilhelmy plate it is advisable to heat the plate in a spirit lamp or Bunsen burner till it becomes red hot. Once cooled to room temperature the platinum late needs to be dipped in piranha solution for an hour post which the plate must be thoroughly washed with distilled water before being used. Also, it must be noted that the plate be very flat and free from undulations. Surface tension of pure water is 72.8 mN/m at 25 °C. Interfacial tension (IFT) between two immiscible

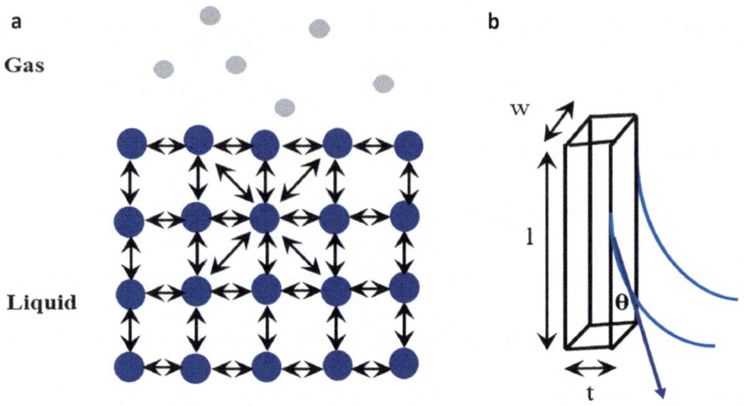

Fig. 4.2 **a** Schematic illustrating the origin of interfacial tension at the liquid–vapor interface, **b** a Wilhelmy plate

liquid phases is an important parameter in many measurements and hence needs to be measured with accuracy. One straight forward and uncomplicated method involves measuring the individual surface tensions of the liquids and then using the relation below:

$$\gamma_{12} = \frac{\Delta W_{12}(\gamma_1 \cos\theta_1 - \gamma_2 \cos\theta_2)}{(\Delta W_1 - \Delta W_2)} \tag{4.2.2.2}$$

where ΔW_{12} corresponds to the weight differential observed when the platinum plate is partially immersed in both the liquids while ΔW_1 and ΔW_2 are the difference in weight observed when immersed partially in the pure liquids 1 and 2 respectively.

When a water-soluble surfactant is dissolved in water, the surface tension of the air–water interface decreases with increasing bulk concentration (C_b) of the surfactant as it gets adsorbed at the air–water interface. This process continues till the interface gets saturated and surface tension reduces to a constant minimum. This value of the surfactant concentration is referred to as the critical micellar concentration (CMC) that is a characteristic of the surfactant. Measurement of surface tension (γ) provides two very important parameters (a) the critical micellar concentration (CMC) and (b) adsorption coefficient (Γ) of surfactants. From thermodynamics, it follows that adsorption coefficient (Γ) which is a measure of the concentration of molecules adsorbed at the interface due to increase in bulk concentration of the surfactant is expressed as

$$\Gamma = \frac{1}{2303\text{RT}} \frac{d(\gamma_{la} \cos\theta)}{d(\log C_b)} \tag{4.2.2.3}$$

where γ_{la} is the liquid (l)–air (a) interfacial tension of the solution containing the surfactant at bulk concentration C_b, θ is the contact angle of the Wilhelmy plate at the liquid–air interface, R is the Universal Gas Constant and T is the temperature in °K [1, 4]. Appendix provides the surface tension of a few common liquids.

4.2.3 *Spinning Drop Tensiometry for Ultralow Interfacial Tension Measurements*

Interfacial tension between oil drops and a detergent solution often attains ultralow values 10^{-3}–10^{-6} mN/m. Such low values cannot be measured with accuracy by the Wilhelmy method. The spinning drop method provides a powerful alternative. The apparatus consists of a thermostated long capillary held horizontally between two supports with one end attached to a high-speed motor that may be rotated at a controlled speed of several thousand (~ 1500 rads/s) as shown in the schematic Fig. 4.3a.

The capillary is filled with liquid ($L1$) having density ρ_1 and a drop of a second liquid ($L2$) of density ρ_2 is introduced in the same capillary. The radius of the drop is measured accurately by a microscope and when the motor is rotated with increasing speeds the droplet elongates along the long axis till the shape of the drop becomes an oblate as shown in schematic Fig. 4.3b. It is observed that the error in measurement is minimal when the ratio of the length (L) of the drop to its diameter (D) exceeds 4. IFT is readily estimated from the relation given as

$$\gamma = \frac{\Delta\rho\omega^2R^2}{4} \tag{4.2.3.1}$$

where $\Delta\rho = \rho_1 - \rho_2$ corresponds to the difference in densities of the two liquids, ω is the angular velocity with which the capillary is rotated and R is the radius of the capillary [1–3].

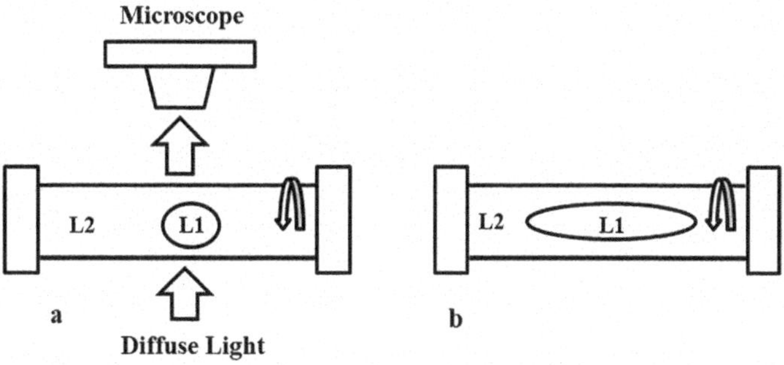

Fig. 4.3 Schematic of spinning drop tensiometer, **a** droplet of oil in the capillary filled with water diameter/length of drop is 1. **b** Ratio of diameter/length of drop is 0.25

4.2.4 Dynamic Surface Tension Measurement by Bubble Pressure Tensiometer

It is well known that surfactants by virtue of their amphiphilic properties get localized at interfaces. Thus when a bubble is formed in a surfactant solution, surfactant molecules rapidly migrate to the newly created interface and the migration process continues till the interface gets saturated. Examining this dynamic process of surfactant migration is important as structure–property correlation of the surfactant with properties like adsorption and diffusion coefficient is easily established which is important in design of products for a specific purpose. A bubble consists of a small volume of air or gas trapped in a bulk liquid. The stability of the bubble is based on the equilibrium of forces, namely the surface tension force at the air–water interface balanced by the internal pressure of the gas entrapped in the bubble and mathematically expressed by Laplace's equation $P_i - P_o = \frac{2\gamma}{R}$ where P_i and P_o corresponds to the gas pressure inside and outside the bubble, γ the surface tension of the liquid–air interface and R the bubble radius. As air is slowly pumped into the liquid the bubble starts to form. Initially, the radius is very large (Fig. 4.4a) that decreases (Fig. 4.4b–d) with increasing volume of the bubble and hence pressure increases. At a radius equal to that of the orifice, pressure in the bubble reaches a maximum post which pressure decreases and drops to zero as the bubble detaches from the tip of the capillary.

Ward and Tordai [5] derived a mathematical relation connecting the concentration of amphiphiles adsorbed at the interface in time (t) that is given by

$$\Gamma(t) = 2C_b\sqrt{\frac{Dt}{\pi}} - 2\sqrt{\frac{D}{\pi}}\int_0^{\sqrt{t}} C_s d(t-\tau) \qquad (4.2.4.1)$$

where C_b and C_s are surfactant concentrations in bulk and at the interface, D is the diffusion coefficient of the amphiphile and τ is a dummy index. A much simpler,

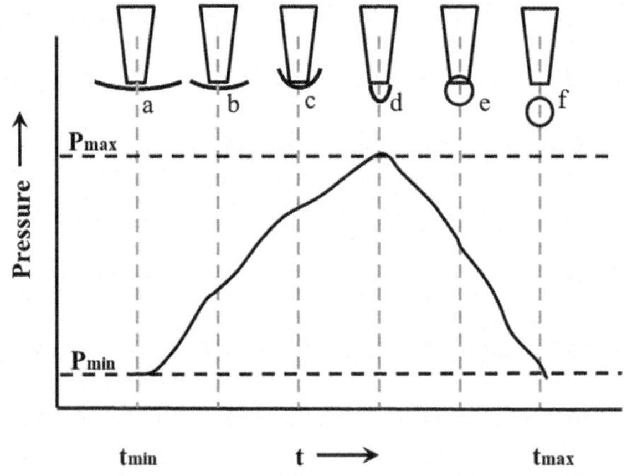

Fig. 4.4 (Top **a**–**f**) shows the formation of bubble at the orifice as a function of time. (Bottom) shows the change in the pressure as a function of time during the process of bubble growth

user-friendly, and frequently used alternative has been provided by Hua and Rosen [6]

$$\gamma(t) = \gamma_m + \frac{\gamma_0 - \gamma_m}{1 + \left(\frac{t}{t^*}\right)^n} \tag{4.2.4.2}$$

where γ_0 is the surface tension of the pure solvent, γ_m is the surface tension of the surfactant solution, t^* and n are dimensionless constants.

4.3 Bulk Properties of Amphiphilic Solutions

4.3.1 Partition Function and Solubilization

A solute is said to be soluble in a solvent if that solute completely disintegrates into its molecular state in that solvent and expressed as gm/lit. Sodium chloride for example readily dissolves in water to give Na^+ and Cl^- ions but is completely insoluble in hexane suggesting that solubility depends on the solute–solvent pair. One well-known parameter used to describe solvency is $\text{Log}_{10}P$ where P is the partition coefficient of the solute in a 1:1 octanol–water mixture. P is evaluated by calculating the ratio of the weights of solute that partitions into octanol and water. While a hydrophilic or ionic molecule will readily dissolve in an aqueous phase, the same will unlikely dissolve in octanol and hence will have a low $\text{Log}_{10}P$ value. On the other hand, a hydrophobic solute will completely solubilize in octanol but will be insoluble in water giving a very low $\text{Log}_{10}P$ value. Hence $\text{Log}_{10}P$ value is a measure of degree of hydrophilicity or hydrophobicity of an organic molecule.

Solubilization [7] is an important parameter as it ensures homogeneity of distribution and hence effectiveness of the product. Increasing temperature of the solvent usually enhances solubility but on cooling the solute invariably precipitates out. An efficient and elegant method of improving solubility of a solute involves adding a suitable surfactant to the solvent (usually water) at a concentration above the critical micellar concentration (CMC) of that surfactant. Above CMC, micelles are formed that have a hydrophobic core that tends to dissolve a hydrophobic solute that is manifested as an enhanced solubility of the solute. In fact, a plot of the solute versus the surfactant confirms above CMC the amount of solute per unit volume of the solvent increases almost linearly.

4.3.2 Electrical Conductivity

Measurement of electrical conductivity is a versatile method of probing into the composition of various materials especially colloidal systems that provides much

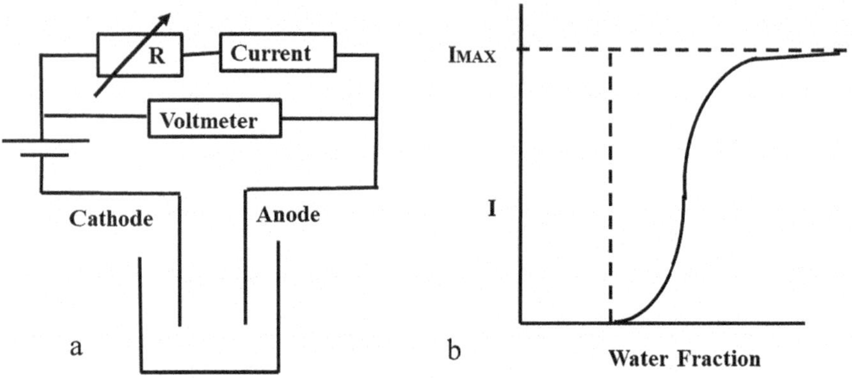

Fig. 4.5 **a** Conductivity cell for measurement of current–voltage characteristics, **b** current versus water fraction in emulsions

insight into the structure–property details of the system. Pure water is non-conducting with a resistivity of 18.2 MΩ-cm and hence unsuitable for examining electrical conductivity. However, adding small amounts of an ionic salt like NaCl or KCl (10^{-6} M) makes it conducting. In an oil in water (O/W) emulsion, oil droplets are stabilized in a continuous water matrix while in a water in oil emulsion (W/O) the reverse is true. When such an emulsion is placed between two conducting electrodes, the conductivity is high for an O/W emulsion while for a W/O emulsion conductivity is usually very low. This feature makes conductivity measurements a powerful tool for examining the nature of an emulsion as shown in the schematic Fig. 4.5. As may be expected, there exists a specific oil to water ratio (ϕ*) above and below which the nature of emulsion are different and this ratio is referred to as the Phase Inversion Concentration (PIC). Similarly, there exists a specific temperature threshold (T*) known as the Phase Inversion Temperature (PIT) at which an O/W changes over to a W/O emulsion and vice-versa.

4.3.3 Zeta Potential

A colloidal system usually consists of sub-micron particles suspended in a liquid. Made right such systems may continue to be stable over very long periods of time that may range from a few months to several years. Detailed studies have established that (a) steric hindrance and (b) electrostatic repulsion between particles prevents aggregation and agglomeration of the particles that make these systems stable. While neutral or particles with low charges tend to agglomerate and precipitate out, highly charged particles indeed make very stable colloidal suspensions. A close inspection of the schematic diagram (Fig. 4.6) shows that while opposite charges strongly adhere to the particle surface forming the Stern layer, charges on the slipping layer contribute to the net charge of the particle. This net charge on the particulate or droplet gives rise to the zeta potential which is a measure of the net charge on the particulate. Detailed theoretical calculations [8] show that the zeta potential (ζ) may be evaluated from the mathematical relation

$$\zeta = \frac{\sigma}{\varepsilon_r \varepsilon_0 a} \tag{4.3.3.1}$$

where (ζ) depends on the ratio of the surface charge (σ) to particle radius a and

$$\sigma = \left[8CN_A \varepsilon_r \varepsilon_0 k_B T 10^3\right]^{1/2} \sinh \frac{Ze\phi_0}{2k_B T} \tag{4.3.3.2}$$

where the relative dielectric constant of the medium (ε_r), permittivity of air (ε_0), and C is the particle concentration in mol/l, T is the temperature in Kelvin, Z is the valency, e is the electronic charge, (ϕ_0) is the potential, N_A is the Avagado's number and k_B is the Boltzmann constant.

4.3.4 Rheology

Rheological properties of materials pertain to their flow properties under externally applied stress or strain [9, 10]. When fluids (air, water, oil, etc.) are subjected to applied stress, these flow obeying the laws of fluid dynamics in accordance with the Navier–Stokes equation. However, materials like gels, semi-solids, colloids, and polymer melts among others exhibit properties intermediate to that of solids and fluids. In general, all fluids may be divided into two classes: (a) Newtonian and (b) Non-Newtonian. Newtonian fluids (water, oils) typically obey the Newton's equation given by

$$\sigma = \eta \left(\frac{\mathrm{d}\gamma}{\mathrm{d}t}\right) \tag{4.3.4.1}$$

that depicts a linear relation between applied stress (σ) and strain rate $\left(\frac{\mathrm{d}\gamma}{\mathrm{d}t}\right)$ and the proportionality constant (η) corresponds to the viscosity of the fluid. Hence a plot of stress versus strain rate for Newtonian fluids is a straight line passing through the origin of the coordinates as shown in Fig. 4.7a.

Non-Newtonian fluids show large deviations from the linear profiles of Newtonian fluids that pass through the origin of coordinates and are characteristics of Newtonian fluids. Bingham fluids for instance also obey a linear stress–strain relationship similar to Newtonian fluids except that it does not pass through the origin but a minimal stress σ_0 is needed to initiate the fluid flow. The stress–strain relation of Bingham's fluid is given as

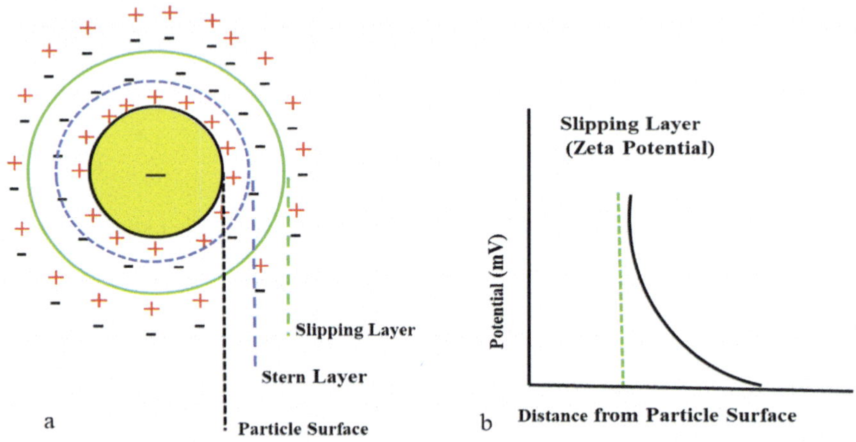

Fig. 4.6 **a** Schematic of zeta potential, **b** zeta potential versus distance from particle surface

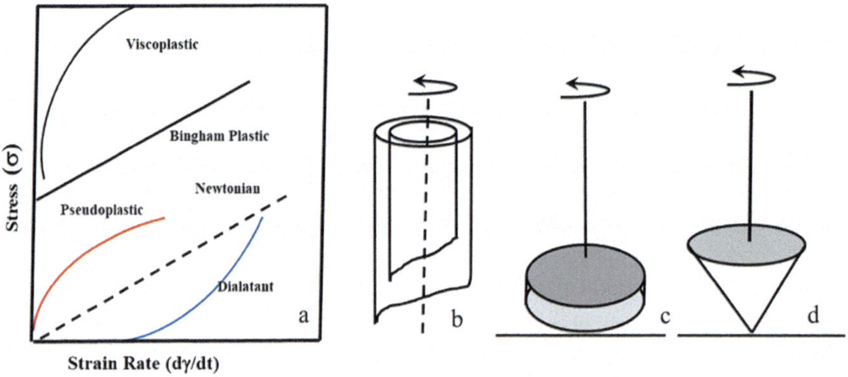

Fig. 4.7 Schematic **a** plot of stress vs strain rate profiles for Newtonian and Non-Newtonian fluids, **b** rotating cylinder rheometer, **c** parallel plate viscometer, **d** cone-plate viscometer

$$\sigma = \sigma_0 + \eta\left(\frac{d\gamma}{dt}\right). \tag{4.3.4.2}$$

Other well-known nonlinear profiles are the following:

$$\text{Ostwald - de Waele equation}\, \sigma = \eta\left(\frac{d\gamma}{dt}\right)^n \tag{4.3.4.3}$$

and

$$\text{Hershel-Bukley equation}\, \sigma = \sigma_0 + \eta\left(\frac{d\gamma}{dt}\right)^n. \tag{4.3.4.4}$$

Although various methods for determining viscosity are known, the simplest and convenient method extensively used is either the rotating cylinder viscometer or the cone-plate viscometer, schematic diagram of which is provided in Fig. 4.7b. The rotating cylinder viscometer consists of two coaxial cylinders with radius R_1 and R_2 where $R_2 > R_1$. The smaller inner cylinder rotates at a fixed angular velocity while the outer larger cylinder remains static. The gap between the outer surface of the smaller cylinder and the inside surface of the outer cylinder is a few millimeters that is filled with material the viscosity of which is being determined. When a torque is applied to the inner cylinder it tends to rotate, but viscous drag opposes that motion and the angular velocity and momentum slows down that is picked up by a series of sensors. This data is processed by a microcomputer that calculates the viscosity using the mathematical relations provided below.

$$\eta = \frac{4M_z(R_2 - R_1)}{\pi(R_1 + R_2)^3 H\omega} \tag{4.3.4.5}$$

$$\tau_{r\theta} \approx \frac{2M_z}{\pi(R_1 + R_2)^2 H} \tag{4.3.4.6}$$

where H is the height of the cylinder, M_z is the moment along the z-direction, ω is the angular velocity of the inner cylinder, $\tau_{r\theta}$ is the torque on the inner cylinder

Combining the above two equations and assuming R $(R_2 - R_1) \ll \frac{R_1+R_2}{2}$ we get

$$\eta = \frac{2(R_2 - R_1)\tau_{r\theta}}{(R_1 + R_2)\omega}. \tag{4.3.4.7}$$

It follows that measurement of the torque $\tau_{r\theta}$ and the angular velocity.

It is worthwhile to note that viscosity is independent of the height of the cylinders but dependent on the gap between the cylinders. In contrast to the rotating cylinder viscometer, the cone-plate rheometer requires a very small amount of sample (5 ml) and hence is a preferred method of determining viscosity. A sinusoidal stress or strain is applied and as discussed the corresponding strain or stress is measured.

Hooke's law is given as $\sigma = G\gamma$ where G is the elastic constant of the material being examined.

Considering a stress of the form

$$\sigma(t) = \sigma_0 e^{i\omega t} \tag{4.3.4.8}$$

then corresponding strain is

$$\gamma(t) = \frac{\sigma_0}{G} e^{i\omega t} \tag{4.3.4.9}$$

implying that strain and stress are in phase for a Hookean material and

$$G = \frac{\sigma_0}{\gamma_0}. \tag{4.3.4.10}$$

For a Newtonian fluid $\sigma = \eta \frac{d\gamma}{dt}$ and it may be readily shown that

$$\gamma(t) = \frac{\sigma_0}{\eta\omega} e^{i(\omega t - \frac{\pi}{2})} \tag{4.3.4.11}$$

implying that strain lags in phase by $\frac{\pi}{2}$.

Over time several models representing viscoelastic materials have evolved, the most common ones being the Maxwell model consisting of an elastic spring representing a pure Hookean solid and a dash-pot containing a Newtonian fluid in series (Fig. 4.8a). The differential equation for Maxwell's model is

$$\frac{d\gamma}{dt} = \frac{1}{G}\frac{d\sigma}{dt} + \frac{\sigma}{\eta} \tag{4.3.4.12}$$

which on solving using the equations $\sigma(t) = \sigma_0 e^{i\omega t}$ and $\gamma(t) = \gamma_0 e^{i(\omega t - \delta)}$ that yields

$$G\prime = \frac{\eta\tau\omega^2}{1 + \omega^2\tau^2} \text{ and } G'' = \frac{\eta\omega}{1 + \omega^2\tau^2} \tag{4.3.4.13}$$

where $\tau = \frac{\eta}{G}$ and $\mathrm{Tan}\delta = \frac{1}{\omega\tau} = \frac{G''}{G\prime}$.

Here *G*' and *G*'' correspond to the elastic and storage modulus respectively. It follows that when $\mathrm{Tan}\delta = 0$.

We have an elastic solid-like behavior and when $\mathrm{Tan}\delta = \infty$ implies a perfect Newtonian liquid.

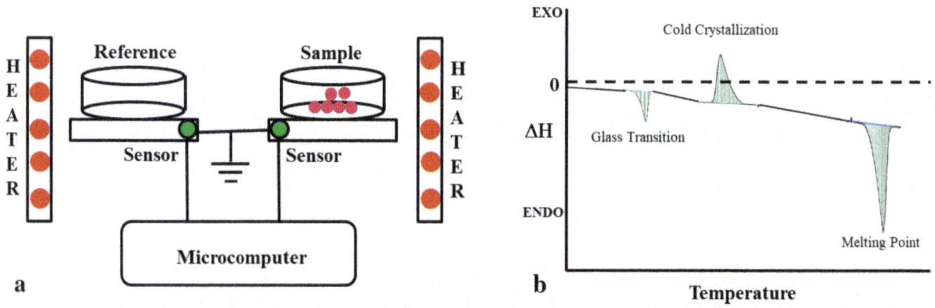

Fig. 4.8 **a** Schematic of a modern differential scanning calorimeter. **b** Schematic of a DSC scan, heat flow versus temperature

4.4 Thermal Analysis

4.4.1 Differential Scanning Calorimetry (DSC)

Differential Scanning Calorimetry (DSC) is a versatile technique that provides considerable insight into the thermal aspects of a material and is based on the simple concept that any material that undergoes structural change or molecular rearrangement either absorbs as heat in events like melting, solidification, or emits heat during crystallization that involves structural phase transformations. The amount of heat absorbed during simple heating or cooling is proportional to the mass of the material. Thus every material is characterized by its unique specific heat capacity (C_p) that is a signature of that material.

In its simplest form, the DSC equipment consists of two identical heaters and temperature sensors that are precision controlled by a computer interfaced with the heaters and sensors as shown in Fig. 4.8a. While one pair of heater and sensor constitutes the reference system, the other pair is the measuring system. Identical containers usually crimpable aluminum pans capable of holding a few milligrams of sample are used. In the sample pan, a known weight of the sample to be examined is placed while the reference pan is kept empty. Both are simultaneously heated at a constant programmed rate and the difference in heat needed to raise both the reference and sample through the same temperature is recorded. A typical DSC plot (Fig. 4.11b) consist of changes in enthalpy $\Delta H (H_{\text{Reference}} - H_{\text{Sample}})$ versus temperature (T) that is recorded digitally.

Such data is invaluable in identifying phase changes occurring in the sample and identifying species that may be formed during heating. In case of melting, heat is absorbed by the sample that is manifested as a negative peak that corresponds to the melting point of that material. The sharpness of the peak profile is often an indicator of the purity of the sample and the order of phase transition it undergoes. Presence of a broad profile with multiple peaks often points to presence of multiple species or phase transitions occurring in the system. A positive ΔH peak corresponds to crystallization while a negative ΔH indicates melting. Thus DSC profiles are excellent tools for examining thermal response of a substance that may be correlated to the changes in its molecular organization during melting, freezing, solidification, and other phase transformations [11].

4.4.2 Thermal Gravimetric Analysis (TGA)

Thermal gravimetric analysis (TGA) involves analyzing change in mass of a material as a function of temperature that provides info on changes to its composition as a function of temperature and thermal stability. Typically, the apparatus consists of a microbalance the temperature of which is precisely controlled by a microcomputer. Figure 4.9a shows a schematic of a TGA instrument and Fig. 4.9b a typical TG

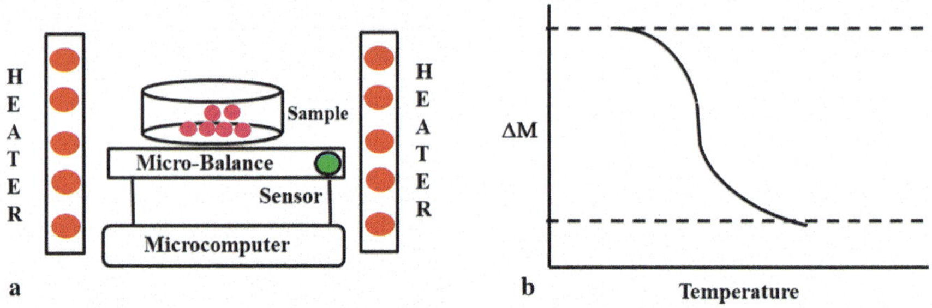

Fig. 4.9 **a** Schematic of a modern thermal gravimetric analyzer, **b** schematic of a TGA, loss in mass versus temperature

profile. As evident from the Fig. 4.9b, the plateau region in the temperature region marked A corresponds to simple heating of the system and no changes to its mass due to evaporation or decomposition. In the region marked B corresponding to the temperature zone $T1$–$T2$ substantial loss of material is observed that suggest loss of volatile components or decomposition of the material and gradient of the line is a measure of rate loss in mass per unit temperature. The plateau again in the region marked D suggests that the material is stable in that temperature zone.

4.5 Imaging Techniques

4.5.1 Optical Microscopy

Optical microscopy [12] is a simple tool routinely used to characterize almost all materials. A schematic diagram of an optical microscope is shown in Fig. 4.10.

The resolution of an optical microscope is usually about 1 μm meaning, two neighboring points separated by 1 μm may be clearly seen as two distinctly separate points while to the naked eye it will appear as a single point. The simple microscope consists of a white light source usually a white LED that is used to illuminate the sample under study. Depending on whether the sample is transparent, translucent as in the case of polymer films or biological samples, the transmission mode is used while for opaque samples as in the case of minerals, the reflection mode is used. A polarizer placed in the path of the incident light and another referred to as the analyzer in the path of the reflected or transmitted light beam from the sample provides insight into the specific alignment of molecular clusters or crystals formed in the sample. Often a xenon or mercury lamp is used for illumination of fluorescent dye labeled samples as in biological samples that help distinguish between different species or binding sites.

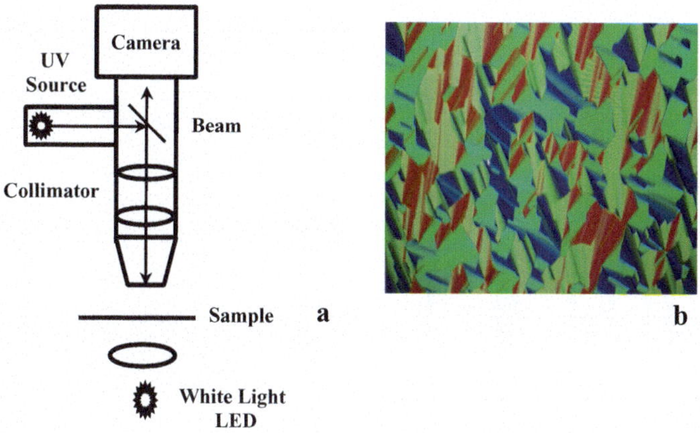

Fig. 4.10 **a** Schematic diagram of an optical microscope with an LED and UV source. LED for use in transmission mode and UV in the reflection mode. **b** Polarized image of a liquid crystal film on glass slide

4.5.2 Fluorescence Imaging of Liquid–Water Interfaces

Fluorescence imaging [13] of liquid–air interface may be achieved by using appropriate dyes doped in lipid layers at the air–water interface. Very small quantities of dye (10^{-3} M) of a water-insoluble fluorescent dye like Nile red, Orange OT, or Sudan III are often used. Alternatively, dye chromophores attached to long lipid or fatty acid chains have been used that get preferentially solubilized in the non-polar hydrophobic phases in the sample that help identify and distinguish between hydrophilic and hydrophobic zones of the monolayer. Illumination of the monolayer may be achieved either in the reflection or transmission mode. Figure 4.11a shows a monolayer of palmitic acid doped with fluorescein palmitate at the air–water interface and illuminated with green light (wavelength 488 nm). Such studies are important to understand the orientation and organization of different ingredients in mixed monolayers that are excellent model systems representing bio-membranes.

4.5.3 Brewster Angle Microscopy (BAM)

Brewster Angle microscopy has emerged as an elegant technique to image non-fluorescent monolayers and aggregates at the air–water interface. The equipment utilizes the simple principle of refraction. It is well known that when a ray of light is incident on the surface of a transparent surface, it gets partly reflected, refracted, and partly absorbed. Absorption is negligible when the surface is thin and transparent. It follows from laws of refraction that there exists an angle θ_c corresponding to which angle between the reflected and the refracted rays is 90° and the relation between this

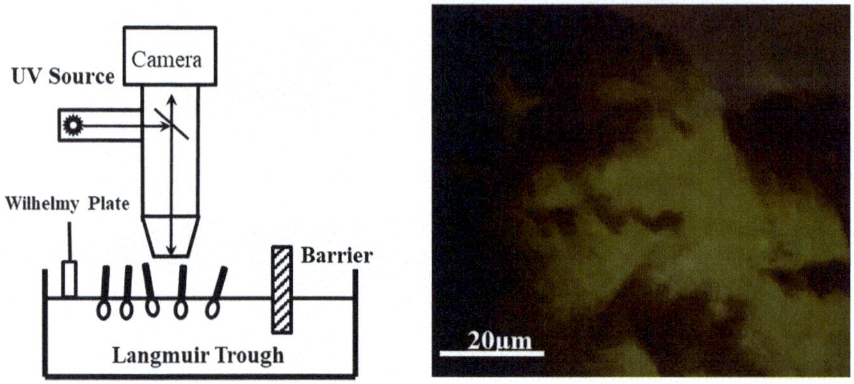

Fig. 4.11 **a** Schematic of fluorescence microscope for imaging air–water interface. **b** Micrograph of a monolayer of 7-Decyl, 2,3 Benzothiazolyl coumarin at the air–water interface in a Langmuir Blodgett Trough. **c** Schematic of Brewster angle microscope

angle and the refractive index (μ) of the refracting medium is given by $\mu = \tan\theta_c$ where μ is the relative refractive index of water with respect to air in this case that yields $\theta_c = 53°$ for the water–air interface as shown in the schematic Fig. 4.12a. A laser beam usually a He–Ne laser is made incident on the air–water interface at an angle of 53° and the reflected beam from the monolayer is imaged using a microscope attached to a CCD camera as shown in Fig. 4.15b. As the monolayer film at the air–water interface is compressed or expanded aggregates are formed in real time that can be imaged with high contrast. Such info is important as it provides insight into the mechanism of formation of aggregates and phase transitions occurring in the monolayer.

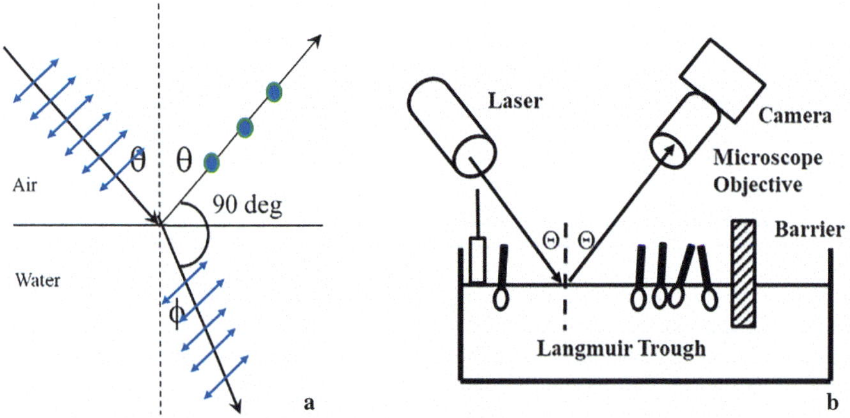

Fig. 4.12 **a** Schematic ray diagram of reflection and refraction occurring at the air–water interface. **b** Schematic of the Brewster angle microscope set up

4.5.4 Scanning Electron Microscopy (SEM)

Scanning electron microscopy (SEM) is a versatile imaging tool that facilitates examining the surface of materials with precision and resolution in the range of 1–20 nm. The equipment consists of four distinct sections (a) an electron gun, (b) electro-optic focusing system (c) condenser system and (d) the detector [14].

The electron gun generates a continuous supply of electrons either through thermionic or field emission from a platinum filament. The electrons produced are accelerated by a pair of anodes maintained at 10 kV and collimated by a condenser system and incident on the sample. Sharp focus and correction for aberrations like astigmatism is achieved by the deflection coils in the objective lens system as shown in Fig. 4.13a. Samples to be examined are usually coated by a thin layer of heavy metal (gold or platinum) to enhance the contrast in the image. When the electron beam impinges on the sample scattered electrons and secondary electrons are obtained from various depths of the sample (Fig. 4.13b) and are collected by the detectors that reconstructs the image of sample. While the secondary and backscattered electrons are used to reconstruct the surface of the sample energy dispersive X-Ray (EDX) detectors are used to identify the composition of samples and in such case the gold coating is not required as it interferes with elemental analysis. It must be pointed out that since the SEM detects scattered electrons it is very important that the sample surface should be uniform and smooth. Because high energy electrons are used to illuminate the samples, intense heat is generated hence ample cooling is a must that is achieved through by cooling the sample with liquid nitrogen.

4.5.5 Transmission Electron Microscopy (TEM)

Transmission electron microscopy (TEM) explores the microstructure of samples that are typically less than 100 nm [15]. This equipment also consists of an electron gun that generates electrons either by thermionic or field emission. These electrons are next accelerated under an applied potential of 100–300 kV and collimated to a narrow beam as shown in Fig. 4.14a. The condenser and electron objective lens system ensures a sharp focus of the electron beam on a small section of the sample to be examined. Here a thin slice of material, usually a few nanometers thick is mounted on a fine copper mesh carbon coated with a nitrocellulose or formvar. A small amount of the material to be examined is placed on the film and dried well. Next, the film is vapor coated with carbon to make it conducting and then exposed to the electron beam. The transmitted beam is next focused on a fluorescent screen allowing direct observing of images. These images are referred to as bright field images. Alternatively, a dark field image is obtained by shielding the transmitted beam with an objective aperture (not shown in the diagram) and the scattered beam from the sample is used to reconstruct the image. Electron diffraction from different sections called SAED (selected area electron diffraction) may be obtained that allows one to

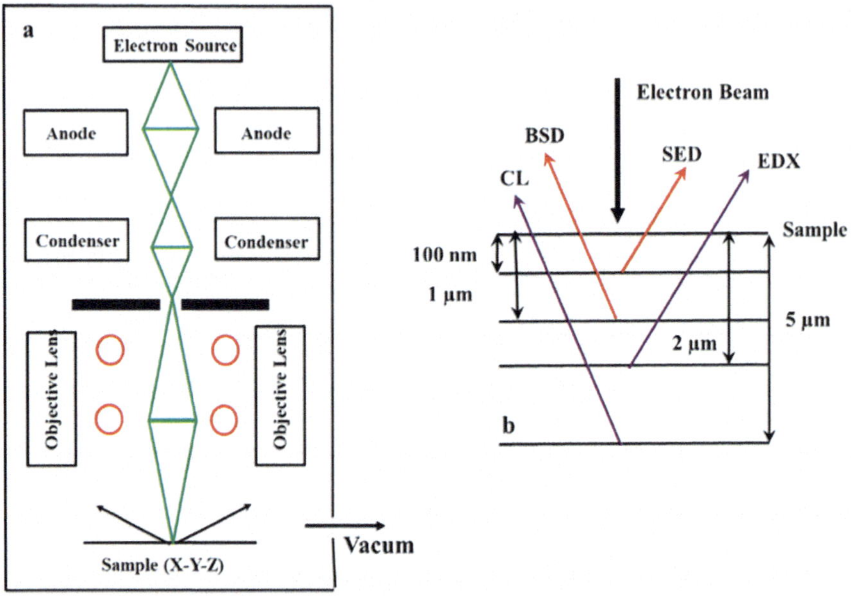

Fig. 4.13 **a** Schematic of a scanning electron microscope, **b** emissions arising from different sample depths. CL—cathode luminescence, BSD—back scattering detector, SED—secondary emission detector, EDX—energy dispersion X-ray detector

conclude whether or not the section of the sample being investigated is amorphous or crystalline. A high angle annular dark field (HAADF) detector is used to construct images from elastically scattered images as shown in schematic Fig. 4.14b.

4.5.6 Atomic Force Microscopy (AFM)

The atomic force microscope (AFM) has evolved as a reliable high-resolution imaging technique [16] that may be used for almost all surfaces conducting and non-conducting, organic and inorganic, charged and neutral, dry or wet, and even delicate biological samples like living cells. A distinct advantage of this technique over others is that it requires minimal effort in sample preparation unlike that required for scanning and transmission electron microscopy where considerable effort and time is needed to appropriately prepare the samples and the quality of the images and its resolution depends largely on the expertise on making samples. Moreover, AFM allows a complete mapping of surfaces in 3D with a resolution of 0.1 nm.

The working principle of an AFM is based on the simple fact that when molecules are separated by several microns the force exerted by one on the other is negligible. However, as one approaches the other and is separated by a few nanometers, the attractive force becomes appreciable and keeps increasing with a decreasing gap

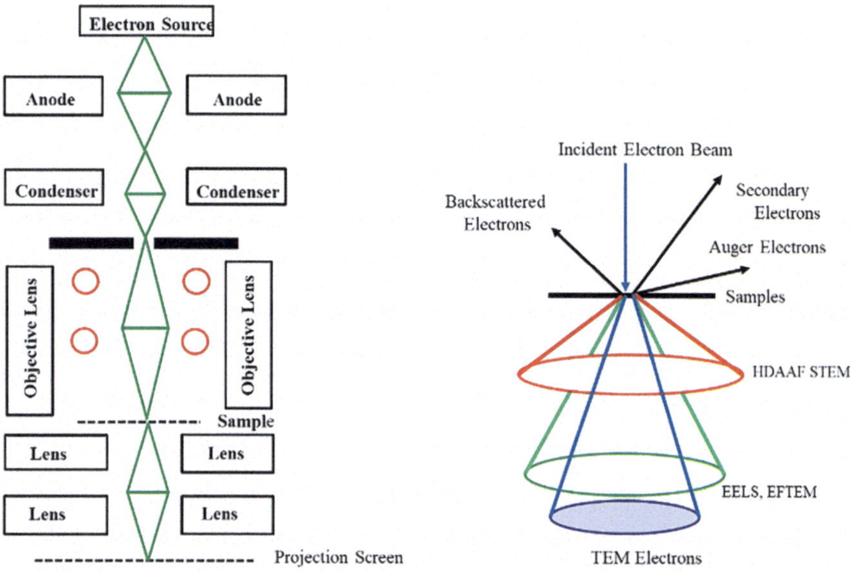

Fig. 4.14 Schematic of a transmission electron microscope

between the two, till it reaches a minimum as shown in Fig. 4.15a. Beyond this point the force is repulsive. The above features shown in the schematic Fig. 4.15a are well established and the mathematical relation governing the force of attraction or repulsion between molecules is expressed as $V(r) = -4\varepsilon\left[\left(\frac{\sigma}{r}\right)^6 - \left(\frac{\sigma}{r}\right)^{12}\right]$ where r is the distance of separation between the surfaces of the two molecules and σ is the distance at which the net potential. $V(r) = 0$ and the minimum value is $V(r) = \varepsilon$ that occurs at $r = 2^{\frac{1}{6}}\sigma$.

The AFM consists of five components namely (a) the probe that consists of a cantilever with a silicon nitride (Si_3N_4) tip, (b) a piezoelectric tube that controls *x*–*y*

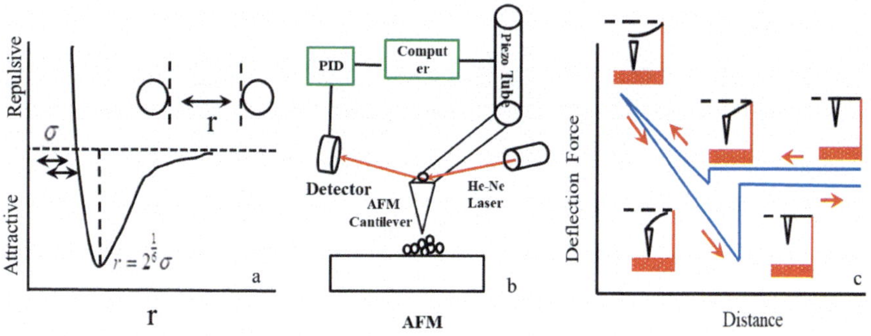

Fig. 4.15 Schematic of **a** Lennard–Jones potential, **b** atomic force microscope (AFM), **c** deflection of the AFM tip as a function of distance from the surface of the sample

movement of the base and the z movement of the tip, (c) position sensitive detector, (d) feedback electronics module and (e) computer and graphics that is depicted in the schematic Fig. 4.15b. Several scanning techniques have emerged over the years namely (a) contact mode, (b) non-contact mode, and (c) tapping mode [12]. In contact mode, the AFM tip is brought in contact with the sample surface and scans the sample surface in the x–y plane. In the non-contact mode, the tip does not touch the surface of the sample but the tip is made to oscillate at its natural frequency of vibration (ν_0) say at a specific height (z_0) from the sample. As the tip encounters a molecule and experiences a force the frequency changes to (ν) say. The difference in frequency ($\nu - \nu_0$) detected is correlated to the distance between the tip and sample and is processed by the computer to yield a 3D topology of the surface. In the tapping mode also known as the intermittent contact mode, the tip is made to vibrate at its resonance frequency while it raster scans the surface of the sample. As the tip moves close to the sample surface (Fig. 4.14c) the increased interaction alters both the amplitude of vibration as well as the frequency of the tip. The change in the amplitude of vibration is used as an input to a feedback loop through a lock-in amplifier that activates a piezoelectric tube to maintain constant distance with the surface as the tip raster scans the surface that generates a 3D topology of the surface. Alternatively, the change in frequency from its resonance value is used to create the 3D topology. Because changes to the frequency is much more sensitive than the amplitude changes, images created using the frequency change as input provides higher resolution images.

Appendix: Physical Properties of Some Common Liquids at 20 °C

Substance	BP (°C) (boiling point)	ρ (gm/cc) density	η (cps) Viscosity	ε dielectric const.	γ (mN/m) surface tension
Water	100	1.00	1	81	72.8
Methanol	64.7	0.791	0.55	32.6(25)	22.7
Ethanol	78	0.816	1.10(25)	24.55(25)	22.1
Butanol	117.7	0.810	2.95	17.80	25
Octanol	196	0.827	10.6(25)	10.30	27.6
Ethyl acetate	77	0.902	0.46	6.02(25)	23.8(25)
Benzene	80	0.874	0.65	2.28	28.8
Hexane	69	0.659	0.31	1.89	18.4
Heptane	98	0.684	0.42	1.92	20.3
Pentane	36	0.813	0.24	1.84	25.5
Cyclohexane	80	0.779	0.98	2.02	25.0

(continued)

(continued)

Substance	BP (°C) (boiling point)	ρ (gm/cc) density	η (cps) Viscosity	ε dielectric const.	γ (mN/m) surface tension
Toluene	110.6	0.865	0.59	2.4(25)	28.5
DMF	153	0.944	0.92	36.7	37.1
DMSO	189	1.10	2.0(25)	46.7	43.5
DMAc	165	0.940	0.92	37.8	36.7
EG	197.3	1.11	18.37	37.0	47.7
Glycerol	290	1.26	11	46.5	64.0

• Terms within parenthesis correspond to temperatures different from 20 °C

References

1. Adamson AW, Gast AP (2012) Physical chemistry of surfaces. Wiley-India Private Limited, New Delhi
2. Kock-Yee L (2015) Surface wetting. Springer, Singapore
3. Masahiko A (2019) Measurement techniques and practices of colloid and interface phenomena. Springer-Nature, Singapore
4. Rosen MJ (2004) Surfactants and interfacial phenomena. Wiley-Interscience, New Jersey, USA
5. Ward AFM, Tordai L (1946) J Chem Phys 14:453
6. Rosen MJ, Hua XY (1982) J Colloid Interface Sci 86:164
7. Shinoda K, Friberg S (1988) Emulsions and solubilisation. Wiley Interscience Publishing, USA
8. Hunter RJ (1988) Zeta potential in colloid sciences. Academic Press, USA
9. Larson RG (1999) The structure and rheology of complex fluids. Oxford University Press, UK
10. Coussot P (2014) Rheophysics : matter in all states. Springer, Switzerland
11. Klimm D (2022) Thermal analysis and thermodynamics. De-Gruyter, Berlin
12. Mertz J (2019) Introduction to optical microscopy. Cambridge University Press, UK
13. Leng Y (2008) Materials characterization: microscopic and spectroscopic methods. Wiley, USA
14. Sultan K (2023) Practical guide to materials characterization, techniques and applications. Wiley, India
15. Reimer L, Kohl H (2008) Transmission electron microscopy: physics of image formation. Springer, USA.
16. Mi L (2023) Atomic Force microscopy for nanoscale physics from single molecules to living cells. Elsevier, USA.

Chapter 5
Surfactants and Self-assembled Structures

5.1 Introduction

Surfactants or surface active molecules are those which when added to a solvent (often water) at low concentrations (0.1% w/w) get rapidly adsorbed at the solvent–air interface thereby altering its interfacial properties, namely surface tension that exhibits a large drop. Surface active molecules are characterized by a hydrophilic (water-loving) and a hydrophobic (water-hating) group built in the same molecule as shown in Fig. 5.1.

Strong adsorption at the solvent-air interface arises from the preference of the hydrophobic group in the molecule to find a hydrophobic microenvironment and likewise for the hydrophilic group to find a hydrophilic microenvironment that gets the surfactant molecule preferentially oriented at the solvent-air interface. The solvent may be polar or non-polar but in this chapter, we will consider the solvent to be water unless specifically mentioned.

Water, despite its seemingly simple molecular structure and abundance, is immensely complex, e.g., its bulk density at 4 °C is the highest while at 0 °C it is lower. Hydrogen bonding between neighboring molecules forms molecular aggregates that range from simple dimers to large clusters. Strong H-bonding (20–35 $kJmol^{-1}$) between water molecules in bulk accounts for its high boiling point (100 °C) and high latent heat of vaporization (40.8 $kJmol^{-1}$). Detailed studies [1, 2] reveal that water clusters exist and are in a state of dynamic equilibrium with their configuration continuously changing as a function of time with an average lifetime (τ) of about 10^{-10}–10^{-12} s and H-bonding extends well over four molecular distances. Therefore, when a surfactant molecule is introduced in water, the hydrophobic group distorts the structural arrangement of water molecules around it thereby increasing the free energy of the system, to counter which the surfactant migrates to the air–water (a/w) interface with its hydrophilic part remaining in contact with bulk water and the hydrophobic end sticking out of the water that is manifested through reduction of surface tension at the air–water interface [3, 4]. In this context, it must be pointed out

A. K. Dutta, *Soft Matter*, https://doi.org/10.1007/978-981-96-0624-5_5

Fig. 5.1 Schematic of a surfactant molecule. Oval shaped blue structure is the hydrophilic head group and the red chain is the hydrophobic chain

Sir Lord Raleigh had demonstrated that when a teaspoon of oil was slowly added to the surface of a lake, it completely covered the surface of the lake and the thickness of the oil layer was found to be a few nanometers that corresponds to the hydrophobic chain length of the oil that confirmed the hydrophobic chain indeed sticks out of the water interface.

5.2 Classification of Surfactants

Surfactants may be broadly classified into four categories (a) anionic, (b) cationic, (c) zwitterionic, and (d) nonionic based on their head group charge.

5.2.1 Anionic Surfactants

Anionics account for about 50% of the total tonnage of surfactants consumed globally. These molecules have a common structure $R\text{-}X^{-}\ M^{+}$ where R is often an alkyl (C_nH_{2n+1}) chain and $n = 8$–16. X is the anionic group (COO^-, OSO_2O^-, POO^- etc.) and M^+ is a cationic usually Na^+. The most commonly used anionics are: (a) Carboxylates: $R\text{-}COO^-\ M^+$, (b) Sulphates: $R\text{-}OSO_2O^-\ M^+$, (c) Sulfonates: $R\text{-}SO_2O^-\ M^+$, (d) Taurates: $R\text{-}CO(CH_2)_2SO_2O^-M^+$ (e) Isenthionates: $R\text{-}CO(OCH_2)_2SO_3^-M^+$ (f) Ethercarboxylates: $R\text{-}(OCH_2CH_2)_nCOO^-M^+$ and (g) Phosphates $R\text{-}POO^-\ M^+$ (Fig. 5.2).

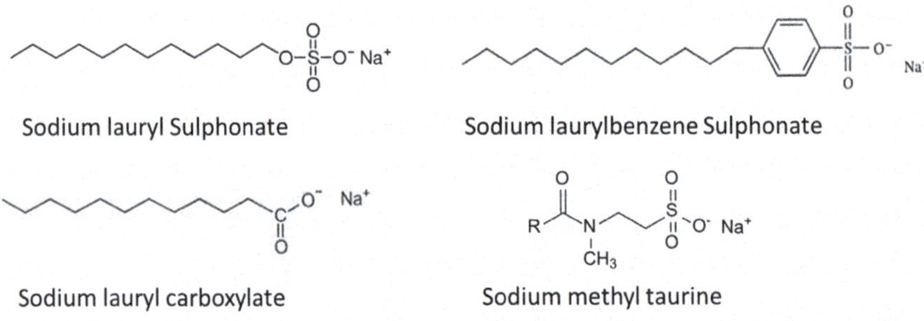

Fig. 5.2 The molecular structure of some common anionics

Carboxylates are soaps used as detergents and cleansing agents in personal care products like body and hand wash [5–7]. Inclusion of EO groups makes anionics milder to the skin and less reactive to sequestering ions. Isenthionates are the mildest of all anionics and hence extensively used in skin and personal care products. Sodium lauryl sulphate (SLS) despite being one of the best-known detergents that are extensively used in shampoos, household detergents, and industrial cleaning products. Being excellent wetting agents these find use in textile, mineral processing, and petroleum industries. Because of their ability to ionize in water, anionic surfactants act as powerful sequestering agents for Ca^{2+} and Mg^{2+} that readily react with the $RCOO^{-}$ group to form salts. In fact, sensitivity of reacting with metal ions is greatest for the carboxylates and lowest for sulfonates. Anionic surfactants form complexes with cationics called "catanionics" that have negligible surface activity and hence anionic–cationic mixtures are incompatible.

5.2.2 *Cationic Surfactants*

These are identical to anionics except that the charge on the head group is positive with the general structure $R_1R_2R_3R_4N^+X^-$ where R's correspond of alkyl, aryl, or alkyl-aryl groups and X corresponds to one of the members of the halogen group. The most extensively used cationic are: (a) Alkylamines: $R_1R_2\text{-}NH_2^+X^-$, (b) Ethoxylated Amines: $R_1R_2\text{-}NH_2^+(C_2H_4O)_2\,X^-$, (c) Alkylimidazolines: $R\text{-}(CH_2)_{n+2}NH_2^+\,X^-$ and (d) Quats: $N^+R_1R_2R_3R_4$. Cationics with one alkyl chain have the highest solubility in water while those with two or more long chains are known to be poorly soluble. However, these are soluble in hydrocarbons [7, 8]. Cationic surfactants are incompatible with anionic surfactants, metasilicates, and phosphates but are compatible with nonionics. In the detergents industry, cationic surfactants are used as fabric softners and in personal care products as hair conditioner and as a mild antimicrobial [6–8]. Cationic surfactants are also used as corrosion inhibitors, flotation aid in mineral procession, antistatic agents, anticaking agents in fertilizers and pigment dispersants (Fig. 5.3).

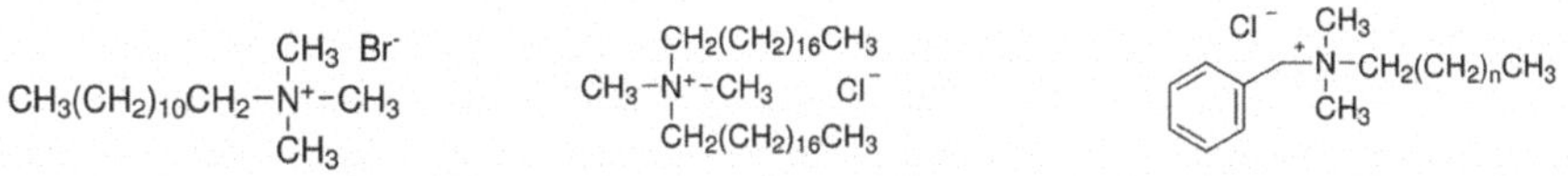

Dodecyl trimethyl ammonium Bromide — Dioctadecyldimethyl ammonium Bromide — Benzalkonium Chloride

Fig. 5.3 The molecular structures of some common cationics

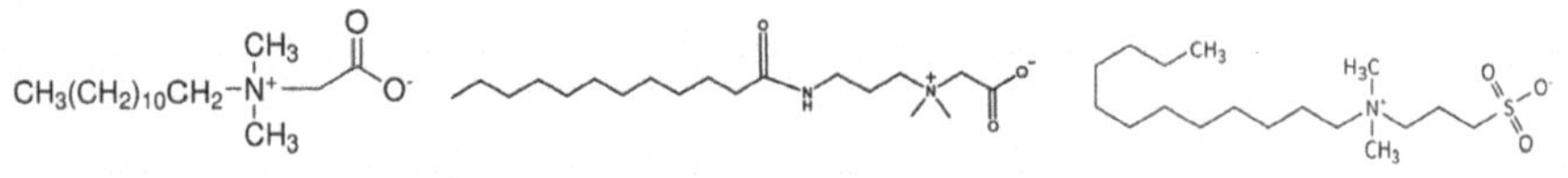

Lauryldimethyl aminoacetate Cocoamidopropylbetaine Lauryl Sultaine

Fig. 5.4 Molecular structure of some common zwitterionics

5.2.3 Zwitterionic or Amphoteric Surfactants

These are electrically neutral but contain both positive and negative moieties separated from each other and located at two different sites on the same molecule, often the end groups. While the positive charge is usually localized on the quartenary ammonium group, the negative charge is localized on a carboxylate, sulphate, or a sulphonate group and because of their large distance of separation between the charges, zwitterionic have large dipole moments. Unlike other surfactants, these may behave as cationic, nonionic, or anionic depending on the pH of its local microenvironment. In an acidic or a low pH medium, the anionic group gets protonated and hence the molecule behaves as a cationic while in an alkaline pH, deprotonation make the molecule behave like an anionic [5, 7] (Fig. 5.4).

Common zwitterionics often used are the following:

(a) Aminocarboxylates: $R\text{-}N^+H_2(CH_2)_nCOO^-$, $R_1R_2N^+H(CH_2)_nCOO^-$ (b) Betaines: $R_1R_2R_3N^+(CH2)_nCOO^-$, (c) Sulfobetaines: $R_1R_2R_3N^+(CH_2)_nCH_2SO_3{}^-$and (d) Amine Oxide: $R\text{-}R_1R_2R_3N^+O^-$ to mention a few. Zwitterionics exhibit anionic or cationic behavior above and below the isoelectric point, at the isoelectric point the same molecules behave like nonionic molecules. Because of their mild characteristic, these are often used in skin care and hair care recipes. Recently, these have been used for corrosion inhibition and enhanced oil recovery [9, 10].

5.2.4 Nonionic Surfactants

Nonionic surfactants are the most extensively used surfactants after anionics. These have "zero" charges on their head groups. Some common nonionic surfactants that are extensively used are: (a) **Alcohol Ethoxylates**: $R\text{-}(OC_2H_4)_n\text{–}OH$, (b), Phenol ethoxylates, (c) **Glycerides**: $R_1\text{-}(CHOR_2)CHOH$, (d) Mono- and Diethanolamides: $R_1R_2NCH_2CH_2OH$ and (e) Polyglucosides. Because of the absence of a charged head group, these are unaffected by pH, presence of electrolytes, and water hardness. These mix freely with all the rest like anionic, cationic, and zwitterionic surfactants. Among all the nonionic surfactants, ethoxylated alcohols are perhaps the most extensively used ingredient across industries that range from detergents, personal care, cosmetics, textiles, paints and agrochemicals. Although nonionics like polyglucosides and sugar esters show an increase in aqueous solubility with increasing temperature like most

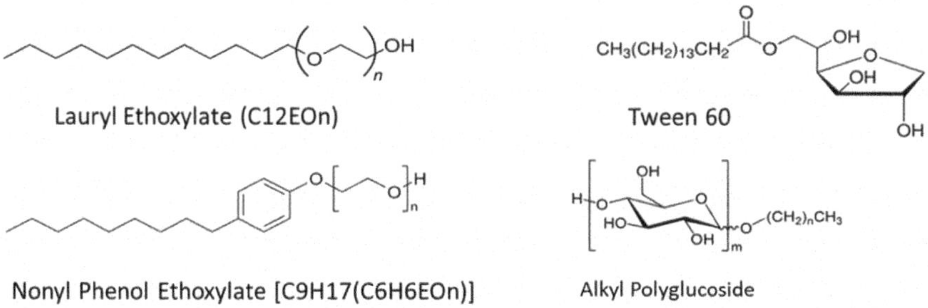

Fig. 5.5 Molecular structure of some common nonionics

materials, alcohol ethoxylates shows a decrease in its solubility in water with increase in temperature. Thus on heating a dilute aqueous solution of an ethoxylated nonionic turns turbid and the temperature at which this occurs is called the cloud point that is an identity of that molecule and its purity. The observed decrease in solubility arises from dehydration of the $-(OC_2H_4)_n$ group with increasing temperature that makes it increasingly oil soluble [5, 7, 11, 12]. Nonionic surfactants are excellent wetting agents and used across industries like, textiles, paper, paint, foods, cosmetics, agrochemicals and detergents (Fig. 5.5).

5.3 Polymeric Surfactants

Polymeric surfactants are polymers that exhibit surfactant-like properties. While the simplest ones could be a polymeric chain attached to a head group that may be either negatively, positively charged, or neutral, other configurations may consist of two or more polymers chemically attached to form a backbone or side chains giving rise to configurations that are referred to in literature as polymeric "combs" or "brushes" [4, 7]. Some of the simplest structures consist of alternate blocks of A and B having configurations like –A–B–A– or $-(A)_n-(B)_m-$ although more configurations are possible. Typical EO-PO copolymers have structures $(EO)_n-(PO)_m-(EO)_n$ and $(PO)_n-(EO)_m-(PO)_n$ are commercially manufactured by BASF under the brand name "Pluronics" and "Reverse Pluronics" that is used extensively in agrochemical formulations, cosmetics and pharmaceuticals. The number of hydrophilic EO and hydrophobic PO groups allow precise control of their HLB making these suitable as wetting, dispersing and stabilizing agents for numerous applications that range from cosmetics, drug delivery, paints, detergents, pharmaceutical and agrochemical formulations. Hypermer-CG6 marketed by ICI consists of a polymethylmethacrylate (PMMA) backbone and polyethyleneoxide (PEO) side chain that is used extensively as a dispersing and stabilizing agent for hydrophobic particles in water. Another extensively used polymeric surfactant is the ATLOX series marketed by Croda. ATLOX-4912 consists of poly(12-hydroxystearic) acid tethered as side chains to a polyethylene glycol backbone. It is insoluble in water having HLB = 6 and molecular weight of about 5000 and used as non-aqueous dispersant and W/O emulsifier.

5.4 Silicone Surfactants

Introduced in the 1950s, these surfactants consist of a siloxane backbone with the dimethylsiloxane (DMS) units end-capped with a trimethylsiloxane (TMS). The hydrophilic group of the molecule may either be an ionic (anionic, cationic, zwitterionic) or nonionic consisting of a short alkyl chain attached to the siloxane backbone. While hydrocarbon surfactants generally lower the surface tension of water to about 30 mN/m, silicone surfactants lower the surface tension further to about 20 mN/m. Silicones have low CMC values and are excellent wetting agents for low-energy surfaces and hence their use in agrochemical, paint, textile industries. Silicone based surfactants have high thermal stability, low toxicity and are non-irritant to human skin and hence used in personal care formulations [13, 14]. Long backbone siloxane compounds and graft copolymers have low CMC's and very low aqueous solubility. However, these are poorly biodegradable and expensive in contrast to their hydrocarbon counterparts which restricts their use.

5.5 Fluorinated Surfactants

Perfluorinated molecules are those in which all C–H bonds in the hydrophobic segment of a surfactant molecule are replaced with C–F while in partially fluorinated surfactants only a fraction of the C–H bonds are replaced. The number of fluorine atoms and their location in the hydrophobic chain cause large changes to the physicochemical characteristics of these molecules in contrast to their hydrocarbon counterparts. The hydrophobic part containing fluorine makes the molecule repel both water and oil. Hence fluorinated surfactants find use in textile and leather coating formulations. Being highly surface active, ppm amounts of fluorinated molecules are sufficient to lower the surface tension of water to below 20 mN/m [14, 15]. Moreover, these are excellent wetting and spreading agents and because these repel both oil and water unlike silicone or hydrophobic surfactants which repel only water these are used in waterproofing and antifogging agents, polishes, paints, textile, surface finishing for paper, and cosmetics. Interestingly, fluorocarbons have low hemolytic activity in contrast to their hydrocarbon counterparts and hence find applications as emulsifiers in foods and drug formulations. The Food and Drug Agency of America (FDA) has approved the use of several fluoro- substituted emulsifiers in food and drugs.

5.6 Gemini Surfactants

These surfactants consist of two identical surfactant molecules separated by a rigid spacer separating the head groups. Three identical molecules attached in a similar configuration are also known. Gemini surfactants have CMCs that are one or two orders lower than their constituent monomers. Moreover, their efficiency in reducing surface tension and interfacial tension is 10–100 times higher than the monomers and they readily form worm-like micelles even in the absence of salts or electrolytes. Unlike the monomers, gemini surfactants at the same concentrations exhibit high viscosity and shear modulus (G') and storage modulus (G'') coefficients. In recent years, these are used in food, personal care [6], enhanced oil recovery [16], and drug delivery systems owing to their excellent, solubilizing, wetting, properties and low Kraft point temperatures [4, 17].

5.7 The Hydrophilic-Lipophilic Balance (HLB)

The hydrophilic–lipophillic (HLB) number is a parameter often used as a measure of hydrophilicty and hydrophobicity of a molecule [4, 7]. The HLB number may vary between 1 and 40 where 1 corresponds to complete solubility in oil and 40 in water, respectively. The resultant HLB of a molecule consisting of different hydrophilic and hydrophobic groups can be estimated from an empirical relation given below:

$$\begin{aligned} \text{HLB} = 7 &+ (\text{sum of all hydrophilic group numbers}) \\ &-(\text{sum of all hydrophobic group numbers}) \end{aligned}$$

A list of various common hydrophilic and hydrophobic groups is listed in Appendix 5.1.

Much work has gone into determination of HLB of mixed surfactants and several empirical relations have evolved, for instance, HLB of phosphatidic alcohol fatty acid esters (sorbitan monoester type) approximates to the relation $\text{HLB} = 20\left(1 - \frac{M_{\text{Hphobic}}}{M_W}\right)$ where $M_{\text{Hydrophobic}}$ corresponds to the molecular weight of the hydrophobic component and M_w is the molecular weight of the entire molecule. Another relation in terms of saponification value (S) of an ester and its fatty acid value (A) is given by $\text{HLB} = 20\left(1 - \frac{S}{A}\right)$. For nonionic surfactants containing only oxyethylene (EO) groups, the HLB may be expressed as $\text{HLB} = \frac{E}{5}$ where E corresponds to the weight percent of oxyethylene in the molecule. For fatty acid esters $\text{HLB} = \frac{E+P}{5}$ where P corresponds to the weight percent of propoxyethylene group. HLB of mixtures is evaluated from the weighted mean of the HLB's and expressed as $\text{HLB}_{\text{Mix}} = \frac{x_1(\text{HLB})_1 + x_2(\text{HLB})_2}{x_1 + x_2}$ where x_1 and x_2 are the mole fractions of the components in the mixture. It must be noted that HLB is empirical in origin but is a good pointer for choosing a surfactant for a given purpose as shown in Table 5.1.

Table 5.1 Correlation between HLB values and its function

HLB range	Function
1–3	Antifoam
3–6	Water in oil Emulsion
7–11	Wetting agent
8–18	Oil in water emulsion
10–15	Detergents
10–18	Solubilizers

5.8 Self-assembly of Surfactants in Solution

When a water-soluble surfactant is added to water in small aliquot the surface tension of water γ decreases with increasing concentration and becomes constant as shown in the schematic Fig. 5.6a. The elbow corresponds to critical micellar concentration (CMC) beyond which surface tension changes very slightly. Other properties like electrical conductivity (σ), and viscosity (η) show sharp changes at the CMC indicating possible structural changes both in bulk and at the air–water (a/w) interface. Considerable research efforts have established that lowering of the surface tension at the a/w interface occurs due to the migration of the surfactant molecules from bulk water to the a/w interface and this process continues till the interface is fully saturated with the surfactant molecules forming an organized monolayer at the interface. Further, increase in the concentration of the surfactant beyond CMC results in the spontaneous formation of highly organized spherical surfactant aggregates with their polar groups at the surface of the sphere and in contact with water while the hydrophobic chains cluster at the center forming a hydrophobic core [4, 5, 7, 18].

These spontaneously formed spherical entities are called micelles. The formation of an organized monolayer at the a/w interface and micelles in bulk arises from

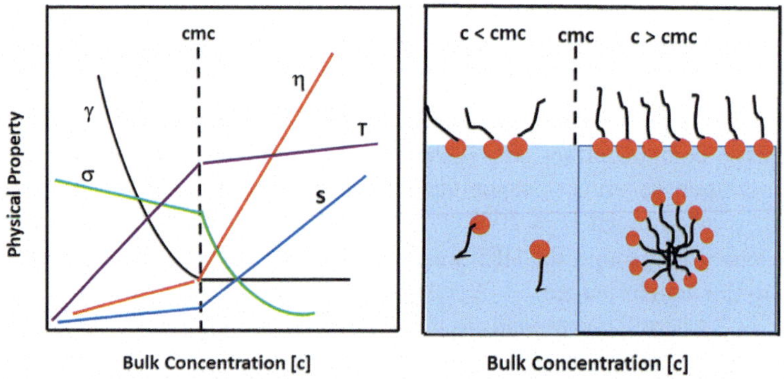

Fig. 5.6 Schematic of **a** physical properties of the system with increasing surfactant bulk concentration **b** organization of surfactants in bulk and at the air-water interface. The symbols used represent the following: surface tension (Υ), conductivity (σ), viscosity (η), turbidity (τ) and solubility (s)

the thermodynamic equilibrium requirement for the system to attain minimum free energy. Detailed studies have confirmed that the equilibrium surface tension γ at the air–water (a/l) interface and CMC depends on (a) molecular structure, (b) the head group, (c) chain length of the surfactant, (d) temperature, (e) polarity of solvent, (f) electrolyte concentration and (g) presence of impurities among other factors [4, 12, 18]. The CMC values of several common surfactants are listed in Appendix 5.2.

Whether or not a material has surface activity is best confirmed experimentally by adding a small amount (0.01% w/w) of the material to pure water and observing changes to its surface tension (if any). Pure water has a constant surface tension of 72.8 mN/m at 25 °C. According to Rosen [4] the efficiency of an amphiphile may be expressed as the minimum concentration (C_{20}) required to lower the surface tension of pure water at 27 °C by 20 mN/m. A quantity similar to pH referred to as pC_{20} ($-\log_{10}[C_{20}]$) is used as a measure of adsorption efficiency of a given surfactant because lowering of surface tension at the air–water interface is due to localization of the surfactant molecule at the interface. As evident from Fig. 5.6a, most physical parameters (conductance, surface tension, osmotic pressure, detergency) does not change significantly above CMC suggesting that once the air–water interface is saturated the above properties become independent of amphiphile concentration. Thus the CMC is a measure of effectiveness of the ingredient in question. According to Rosen [4] the ratio of CMC/C_{20} is a measure of the ratio of "effectiveness" to "efficiency" that is an indicator of whether micellization or adsorption is a dominant process in that system [12].

As surfactants consist of long hydrophobic chains attached to a hydrophilic head group, it follows that with increasing chain length, solubility will decrease at any specific temperature but will increase with increasing temperatures for a given chain length. Similarly, CMC will also increase with temperature. Kraft Temperature, named after Sir Friedrich Kraft corresponds to the minimum temperature at which micelles are formed, i.e. solubility of amphiphiles at Kraft temperature is equal to CMC at that temperature [4, 7] and corresponds to the intersection of the two curves as shown in Fig. 5.7.

Fig. 5.7 Schematic of the variation of solubility (black) and CMC (red) of a surfactant with temperature

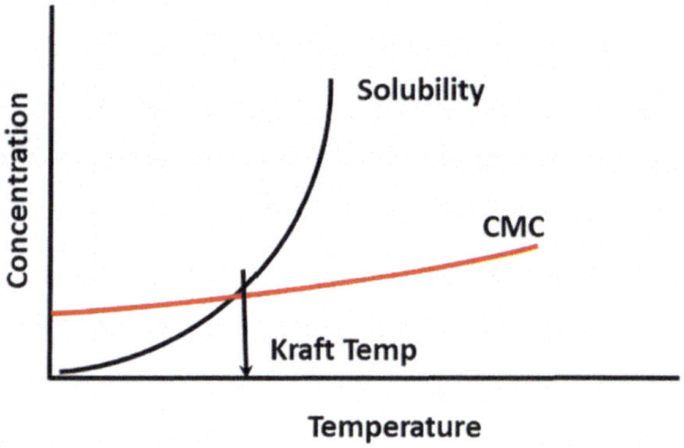

Surfactant molecules tend to aggregate through hydrophobic-hydrophobic interactions and get preferentially organized to meet thermodynamic minimum free energy requirements. As hydrophobic groups do not find a favorable environment in a polar solvent like water, they tend to move away to the water–air interface and reside with their hydrophobic tail sticking out of water at the a/w interface. With increasing surfactant concentration, the air–water interface first gets saturated post which micelles start forming spontaneously. These micelles are aggregates of surfactant molecules forming a spherical structure with the hydrophobic groups forming a core water as shown in Fig. 5.1b. Using fluorescent probes to probe the polarity of the hydrophobic core, it has been concluded that the hydrophobic core has polarity corresponding to that of methanol–water mixture [19]. Detailed studies have confirmed that the shape, size, and number of surfactant molecules that form the micelle also called aggregation number depends on the molecular structure and solvent system used. Different models [4, 7] have been proposed for the formation of micelles with each one having its own set of advantages and disadvantages. However, the simplest one is the mass action where the initial state at concentrations below CMC contain monomers that are randomly oriented in solution which aggregate together to form micelles following the reaction $nA \Leftrightarrow A_n$ and the rate of reaction is given by $k = \frac{[A_n]}{[A]^n}$ that yields the Gibbs free energy change as $\Delta G^\circ = -\mathrm{RT}\ln K \approx \mathrm{RT}\ln \mathrm{CMC}$.

5.9 Critical Packing Parameters (CPP)

Considering the micelle to be a spherical entity, the aggregation number N may be expressed as $N = \frac{V_{\mathrm{micelle}}}{V_H} = \frac{4}{3}\pi\frac{R^3_{\mathrm{micelle}}}{V_H}$ where V_H is the volume of a hydrophobic chain of a single molecule, R_{micelle} is the radius of the micelle. Again in terms of the surface area, the expression for aggregation number is $N = \frac{\mathrm{Area}_{\mathrm{micelle}}}{a} = 4\pi\frac{R^2_{\mathrm{micelle}}}{a}$ that yields $\frac{V_H}{R_{\mathrm{mic}}a} = \frac{1}{3}$ where the ratio $\frac{V_H}{R_{\mathrm{mic}}a}$ is called the critical packing parameter (CPP), a is the surface area of the polar head group of a single molecule. Because the radius of the spherical micelle cannot exceed the maximum alkyl chain length (l_{max}) we get $\mathrm{CPP} = \frac{V_H}{l_{\mathrm{max}}a} \leq \frac{1}{3}$ for a spherical micelle. Table 5.2 provides a list of CPP values corresponding to different structures.

Table 5.2 Correlation between Critical Packing Parameter (CPP) and various organized structures formed in different phases

CPP (p)	Structure	Phase
$p < 1/3$	Spherical	L1
$1/3 < p < 1/2$	Cylindrical	Hexagonal (H1)
$1/2 < p < 1$	Lamellae	Lamellar (Lα)
$1 < p < 2$	Bilayer	
$2 < p < 3$	Inverse cylindrical	Hexagonal (H2)
$P > 3$	Inverse spherical	L2

It must be pointed out that the shape and size of micelles depend on their molecular structure and other factors like charge on their head group, presence of co-solutes, co-solvents, and electrolytes and polymers that form complexes that can produce completely different structures [4, 7, 20, 21]. Sodium dodecyl sulphate (SDS) for example has a large head group that gets ionized in water. While at low salt concentration (0.1 M NaCl) it forms spherical micelles, at higher salt concentrations it forms cylindrical micelles as presence of counterions reduces the effective head group charges hence '*a*' decreases and the packing parameter p increases beyond 1/3 hence spherical micelles are not formed instead cylindrical micelles are created. Similarly, double-chained surfactants have a large volume and hence value of $p \sim 1$ is leads to the formation of vesicles or lamellae and perhaps sponge phases. Indeed, phosphatidylethonalamine, phospatidylserine in high salt concentrations form planar bilayers while double chained lipids, e.g. phosphatidylcholine, phosphatidyl inositol, and dialkyldimethyl ammonium salts with large head groups form flexible bilayers or vesicles. In contrast, double-chained lipids with small head groups, and nonionic lipids, e.g., phosphatidylethonalamine form inverse micelles.

Triple-chain surfactants for similar reasons often form bilayers.

5.10 Concentration and Temperature Dependent Phases

At low surfactant concentrations below CMC, only monomers exist and at concentrations above CMC, micelles are formed. These micelles are aggregates of surfactant molecules closely packed together as small, spherical entities. The size and shape depend on the molecular structure of the surfactant and the aggregation number (N) which is the total number of monomers that constitute the micelle. Aggregation number depends on the molecular structure of the surfactant and on the ionic strength and nature of ions present in the solution. With increasing surfactant concentration, and nature of electrolyte, and its concentration different structural configurations like cylindrical and wormlike micelles get formed [20–23]. Because aggregation of surfactant monomers occurs as a result of the sum total of attractive (van der Waals, hydrogen bonding, hydrophobic-hydrophobic) and repulsive (electrostatic) interactions, the physical properties of the aggregates lie between that of pure liquids and crystalline solids and hence referred to as liquid crystalline (LC) phases. Liquid crystals are of two types: (a) thermotropic and (b) lyotropic. The physical properties of thermotropic LC materials vary with temperature while in lyotropic materials the properties are concentration-dependent.

Figure 5.8 shows the phase transition trend of common surfactants with increasing surfactant concentrations. Figure 5.3. Schematic of various phases as a function of concentration and temperature. For surfactants with HLB (40–10), at low concentrations (below CMC) the solutions are isotropic followed by micellar state at CMC. Further increase leads to the formation of the hexagonal (H1) and cubic (V1) phases. Further increase leads to inverse cubic (V2) and hexagonal (H2) phases followed

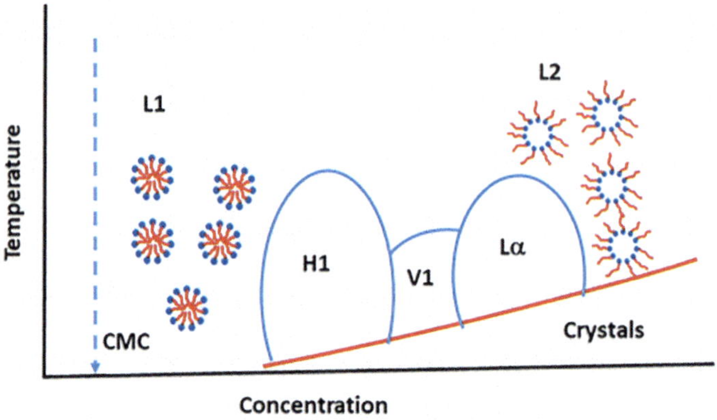

Fig. 5.8 Phases as a function of temperature and concentration

by inverse micellar and isotropic phases once again. The inverse phases can be reproduced when the surfactants have low HLB (1–10).

Appendix 5.1: Critical Micellar Concentration (CMC) of Some Common Surfactants

Surfactant	Type	CMC(mM)
Sodium dodecyl sulphate	Anionic	8.3
Sodium decyl sulphate	Anionic	3.3
Sodium octyl Sulphate	Anionic	1.33
Sodium p-dodecyl benzene sulphonate	Anionic	1.20
Dodecylammonium chloride	Cationic	14.7
Dodecyltrimethylammonium chloride	Cationic	20.3
Dodecylammonium bromide	Cationic	65.0
Dodecyltrimethylammonium bromide	Cationic	15.6
C10E8	Nonionic	1.0
C12E8	Nonionic	0.071
C14E8	Nonionic	0.090

Reference [4]

Appendix 5.2: HLB of Different Groups

Hydrophilic group	HLB number
$-SO_4Na$	35.7
–COOK	21.1
–COONa	19.1
–COOH	2.1
–OH (free)	1.9
–OH (sorbitan ring)	0.5
–N (tertiary amine)	9.4
–O	1.3

HLB = 7+ Sum total of all (Hydrophilic + Hydrophobic) group numbers. Reference [4]

References

1. Frank HS, Wen WY (1957) Discuss Faraday Soc 24:133–140
2. Frank F (1983) Water. Royal Society of Chemistry, London
3. Jones MN, Chapman D (1995) Micelles, monolayers and biomembranes. Wiley-Liss Inc., N.Y.
4. Holmberg K, Jonsson B, Kronberg B, Lindman B (2003) Surfactants and polymers in aqueous solutions, 2nd edn. Wiley, N.Y.
5. Farn RJ (2006) Chemistry and technology of surfactants. Blackwell Publishing, U.K.
6. Rhein LB, Schlossman M, O'lenick A, Somasundaran P (2007) Surfactants in personal care products & decorative cosmetics. CRC Press, N.Y.
7. Tadros TF (2005) Applied surfactants: principles and applications. Wiley-VCH, N.Y.
8. Goncalves RA, Holmberg K, Lindman B (2023) J Molecular Liquids 375:121335
9. Brycki B, Szulc A (2021) J Molecular Liquids 344:117686
10. Pal N, Hoteit H, Mandal A (2021) J Molecular Liquids 339:116811
11. Schick MJ (1987) Nonionic surfactants-physical chemistry. Marcell-Dekker, N.Y.
12. Rosen MJ (1989) Surfactants and interfacial phenomena, 2nd edn. Wiley Interscience, N.Y.
13. Hill RM (1999) Silicone surfactants. Marcell-Dekker, N.Y.
14. Tan YF, Xiong XM, He ZY, Cao F, Sun DS (2019) J Phys Chem 123:1390
15. Kissa E (2001) Fluorinated surfactants & repellents, 2nd edn. Marcell-Dekker, N.Y.
16. Hussain SMS, Adewunmi AA, Mahboob A, Murtaza M, Zhou X, Kamal MS (2022) Adv Colloid Interface Sci 303:102634
17. Ospenikov AS, Gavrilov AA, Artykulnyi OP, Kulkin AI, Novikov VV, Slibaev AV, Phillippova OE (2021) J Colloid Interface Sci 602:590
18. Evans DF, Wennerstorm H (1994) The colloidal domain: where physics, chemistry, biology and technology meet. VCH, N.Y.
19. Kalyansundaram K (1991) Photochemistry in organized and constrained media. In: Ramamurthy (ed). VCH Publishers, N.Y.
20. Raghavan SR, Feng Y (2017) Wormlike micelles: solutions, gels or both. In: Dreiss CA, Feng Y (eds) Wormlike micelles: advance in systems, characterization and applications. Royal Society of Chemistry, U.K.
21. Kamal MS (2016) J Surfactants Detergents 19(2):223–226

22. Ghosh S, Ray A, Pramanik N (2020) Biophys Chem 265:106249
23. Eastoe J (2010) Surfactant aggregation and adsorption at interfaces, 2nd edn. In: Cosgrove T (ed) Colloid science principles: methods and applications. Wiley, U.K.

Chapter 6
Emulsions

6.1 Introduction

The immiscibility of oil and water has been known for centuries and attempts to mix them have always resulted in the formation of two distinct layers. However, surface active agents often referred to as surfactants or emulsifiers when added to oil and water result in a milky white liquid where the layer of oil and water does not separate out immediately but may continue to be in that state for several hours or days. This milky liquid is called an emulsion and is actually a liquid–liquid dispersion as may be confirmed by a simple optical microscopic inspection of the milky liquid. When oil is the dispersed phase and water is the bulk continuous medium, the emulsion is an oil-in-water (O/W) emulsion. Similarly, when the dispersed phase is water and the bulk phase is oil we have a water-in-oil (W/O) emulsion. The oil phase need not necessarily be a conventional oil like paraffin or vegetable oil but may also be a non-polar hydrocarbon e.g. benzene, hexane, pentanol, dichloroethane while the water phase may be replaced by a polar liquid e.g. ethanol, polyethylene glycol. Several food items such as milk, butter, mayonnaise, sauces, and even ice cream consist of micron-sized oil or fat droplets dispersed in an aqueous base are classic examples of emulsions that we come across in our daily lives.

Over decades, various terminologies like "coarse", "macro", "mini", "micro", and "nano" emulsions have emerged that are based on their physical appearance, stability, and droplet size [1, 2]. Microemulsions are transparent and have droplet size (d_{50}) less than 100 nm. Mini- or Nano-emulsions have (d_{50}) between 100 nm to 1 μm and macroemulsions have (d_{50}) greater than 1 μm. While microemulsions are thermodynamically stable, all others are unstable and hence over time separates out as oil and water layers.

In this chapter, we examine in brief the physico-chemical properties of macroemulsions, process of making it and the various factors that stabilizes or destabilizes these emulsions.

A. K. Dutta, *Soft Matter*, https://doi.org/10.1007/978-981-96-0624-5_6

6.2 Emulsion Types

Emulsions may be classified as:

(1) Emulsions (a) Oil in water (O/W) and (b) Water in oil (W/O) as shown in the schematic Figs. 6.1a–d, and 6.2. Multiple emulsions (a) Oil in water in oil (O/W/O) and (b) Water in oil in water (W/O/W) as shown in the schematic Fig. 6.1e–f.

6.2.1 Optical and Fluorescence Microscopy

Emulsions, whether oil in water (O/W) or vice-versa, appear milky white to the naked eye. This is due to scattering of white light by the droplets of oil or water in the O/W or W/O emulsions. To identify the nature of the emulsion a small amount (0.002 M) of an oil-soluble fluorescent aromatic hydrocarbon (perylene) was added to the emulsion and thoroughly agitated. Next, a drop of this dye-treated emulsion is

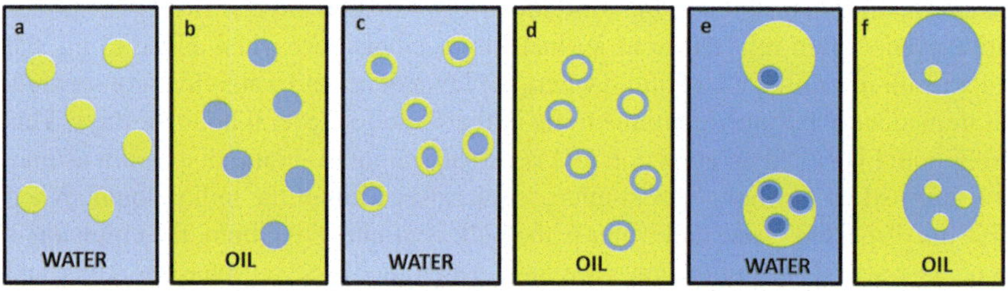

Fig. 6.1 Schematic of **a** oil in water O/W emulsion, **b** water in oil W/O, **c** water in oil in water W/O/W emulsion, **d** oil in water in oil O/W/O emulsion, **e** multiple emulsions of W/O/W and **f** multiple emulsions of O/W/O. Blue indicates water and Yellow indicates oil

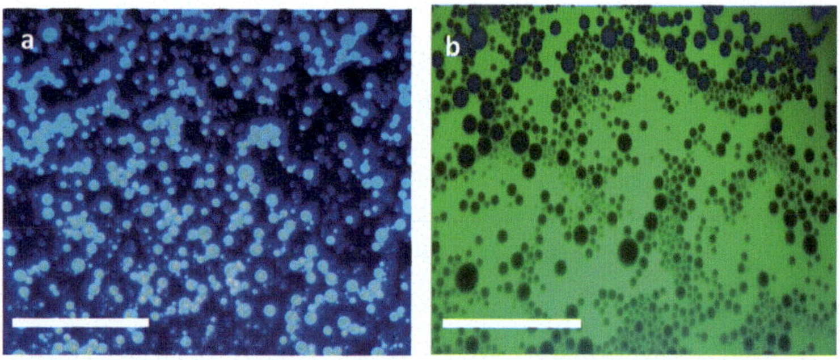

Fig. 6.2 **a** Fluorescence micrograph of paraffin oil in water emulsion. Perylene (0.002 M) was used to label paraffin droplets and illuminated with blue radiation from a xenon arc lamp using a 400 nm filter. **b** Bright field image of a soya oil emulsion to which sodium fluorescein (0.01 M) was used to label the aqueous portion of the emulsion. White bar represents 20 μm

observed through a fluorescence microscope. The dark field image (Fig. 6.2a) shows intense blue-colored oil droplets on a dark background confirming that it is an o/w emulsion. On the other hand, when sodium fluorescein was added to an emulsion and observed under white light in the transmission mode, the oil droplets appeared as dark dots on a green background as shown in Fig. 6.2b.

6.2.2 Electrical Conductivity

Electrical conductivity measurements are simple yet versatile means of determining the nature of an emulsion. Oils are non-conducting and hence a water-in-oil (W/O) emulsion placed between two electrodes across which an applied voltage has been applied will exhibit low conductivity. On the other hand, oil in water (O/W) emulsion shows high conductivity. It must be pointed out that pure water is non-conducting hence a small amount (1×10^{-6} M) of an electrolyte like NaCl or KCl is added when evaluating an emulsion. The relation [3] connecting the volume fraction of the dispersed phase ϕ_d with the conductivity [3] of the emulsion (σ_e), continuous phase (σ_c), and the dispersed phase (σ_d) is given by

$$\phi_d = \left(\frac{\sigma_e - \sigma_c}{\sigma_e + 2\sigma_c}\right)\left(\frac{\sigma_d + 2\sigma_c}{\sigma_d - \sigma_c}\right). \tag{6.2.2.1}$$

6.3 Thermodynamics of Emulsification

Let us consider an oil droplet of radius R_0 and bulk density ρ introduced in water and the oil–water interfacial tension is γ_{12} where suffix "*o*" stands for oil and "*w*" for water. Then Gibbs free energy (ΔG) of the system is

$$\Delta G = \Delta A \gamma_{0w} - T\Delta S \tag{6.3.1}$$

where G is the Gibbs free energy, T is the temperature in °K, S is the entropy of the system, and A is the surface area of the drop. It can be readily shown that if the above spherical drop of radius R_0 is split into N droplets of radius R, then R can be related to R_0 by the relation $R = N^{-1/3}R_0$ considering conservation of mass and the corresponding change in total surface area is

$$\Delta A = 4\pi R_0^2(N^{1/3} - 1) \tag{6.3.2}$$

where ΔA is the change in the total surface area produced by breaking up of a single drop into N smaller droplets. Plugging this value ΔA into the expression for Gibbs free energy and neglecting the $T\Delta S$ term which is actually a small positive quantity,

Table 6.1 Compares the physical properties and preparation process of various emulsion types

Type of emulsion	Appearance	d_{50} (nm)	ΔG	Preparing process
Nanoemulsion	Transparent or translucent	< 100	> 0	High energy input needed
Miniemulsion	Translucent	100–1000	> 0	Energy input needed
Microemulsion	Transparent clear	< 100	≤ 0	Spontaneous
Macroemulsion	Milky and opaque	> 1000	> 0	Energy input needed

*d_{50} corresponds to the average droplet size in an emulsion

we get

$$\Delta G = 4\pi \gamma_{ow} R_0^2 (N^{1/3} - 1). \tag{6.3.3}$$

Clearly, $\Delta G > 0$ as $N^{1/3}$ is a large number and hence energy will be needed to be supplied from an external source to the oil–water system to split the oil drop into smaller droplets as required for macro- and mini-/nano-emulsions. It also follows that smaller the droplet, larger is the corresponding value of $N^{1/3}$ hence larger is the magnitude of energy needed. Adding a suitable surfactant can lower γ_{ow} to very low values (10^{-3} mN/m) that makes $\Delta G \approx 0$ implying that in the presence of a suitable emulsifier, the oil drop will spontaneously emulsify and would be extremely stable which is indeed the case for microemulsions [1, 4] (Table 6.1).

6.4 Internal Structure of Emulsions

As mentioned above, electrical conductivity measurements offer a unique means of exploring the internal structure of emulsions. It has been observed that in an emulsion of paraffin, water and a nonionic [4] conductivity increases with the water volume fraction. It was observed that at a very volume fraction of water ($\phi_W < 0.1$) the conducting current was very low, but on increasing the water fraction slowly it was found that there exists a critical volume fraction of water ($\phi_w = \phi_c$) above which current rapidly increases and conductivity reaches saturation as shown in the schematic (Fig. 6.3). These observations have been explained by various authors in light of the percolation theory [5, 6].

At low water fractions ($\phi_w < \phi_c$), electrical conductivity (σ) is low suggesting that water pools are separated by large stretches of oil that prevent the ions from one electrode from reaching the other. At the threshold ($\phi_w = \phi_c$), the above-mentioned water pools interconnects to form channels connecting the electrodes that make conduction possible. Beyond this threshold, with increasing water fraction, conduction rapidly increases as larger connectivity is established through increasing number of pathways till a maximum is attained. A plot of the normalized conductivity or current versus the water volume fraction (ϕ_w) is represented by a sigmoidal curve as shown in the schematic Fig. 6.2b. According to the percolation theory [5–8], the normalized

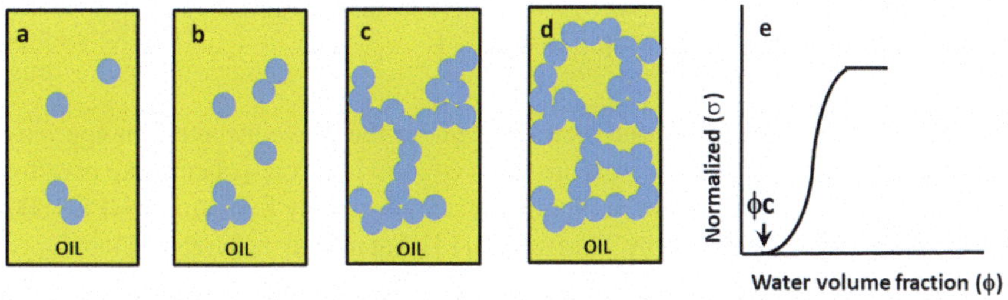

Fig. 6.3 **a–d** Schematic illustrates the development of conducting channels with increasing volume fraction of water in the W/O emulsion. Blue represents water and Yellow represents oil. **e** Shows a typical sigmoidal curve of normalized conductivity and water fraction (ϕ) that is characteristic of a percolation process

current may be expressed as

$$i = |\phi - \phi_c|^t \quad \text{for} \quad \phi > \phi_c \tag{6.4.1}$$

and

$$i = |\phi_c - \phi|^{-s} \quad \text{for} \quad \phi < \phi_c \tag{6.4.2}$$

where the exponents t and s are the scaling exponents. Detailed theoretical and experimental studies have confirmed that these exponents can be correlated to the fractal dimensionality of ion conduction in the emulsion system [6, 7].

6.5 Rheology of Emulsions

The rheological properties of emulsions are of considerable importance in the food and cosmetics industry as they cater to several consumer needs and aspirations. Considering hard spherical particles dispersed in a Newtonian fluid, Einstein derived a theoretical relation connecting relative viscosity and phase volume that is given by $\eta_r = 1 + 2.5\phi$ where (η_r) is the relative viscosity and (ϕ_d) is the dispersed phase volume fraction. It has been demonstrated through various works that the above relation holds good for low ϕ_d typically $\phi < 0.01$. An important feature of Einstein's equation is that it is a linear relation and independent of the characteristics of the materials constituting the emulsion or temperature which is surprising. Brinkman and Roscoe [8, 9] proposed an equation $\eta = \eta_0(1 - \phi_d)^{-2.5}$ that modifies to $\eta_r = 1 + 2.5\phi_d + 14.1\phi_d^2$ has been shown to hold good when $\phi_d < 0.06$. While the first term corresponds to that of Einstein's equation, the second term originates from hydrodynamic interactions between the droplets that become prominent at a higher volume fraction of the dispersed phase. Considering the dispersed phase to be made of spherical hard spheres Kreiger and Douharty [10] proposed

$$\eta_r = \eta_0 \left(1 - \frac{\phi_d}{\phi_{\max}}\right)^{-2.5\phi_{\max}}. \tag{6.5.1}$$

Expanding the expression as a series and under the low concentration approximation the expression simplifies to Einstein's equation. In real systems like paraffin oil–water emulsions addition of nonionic surfactants namely Synperonic NPE 1800 and its analogs were examined by Tadros et al. [11] that indeed the Kreiger-Dougharty relation η_r and ϕ holds good.

Furthermore,

$$\eta_r = \left(1 - \frac{\phi}{\phi_{\max}}\right)^{-|\eta|\phi_{\max}} \tag{6.5.2}$$

where $|\eta|$ is the intrinsic viscosity and $\phi_{\max}$ is the maximum packing fraction [12]. At high-volume fractions, $\phi \geq 0.6$ the viscosity (η) varies inversely as the diameter (d_m) of the droplet and the distance of separation (x_m) between the droplets is given by

$$x_m = d_m \left(\left(\frac{\phi_{\max}}{\phi}\right)^{\frac{1}{2}} - 1\right) \tag{6.5.3}$$

$|\eta|$ is numerically equal to 2.5 while $\phi_{\max}$ is the maximum packing fraction and is equal to 0.64 for random packing and 0.74 for hexagonal close packing of monodisperse spheres. For polydisperse systems, $\phi_{\max}$ is found to exceed the above theoretical values and several empirical relations have been proposed.

For low shear and concentrated dispersions, a good fit is often obtained using the relation

$$\eta_r = \frac{1 + 1.5\phi_d(1 + \phi_d - 0.189\phi_d^2)}{1 - \phi_d(1 + \phi_d - 0.189\phi_d^2)}. \tag{6.5.4}$$

Another expression due to Thomas [13] is

$$\eta_r = 1 + 2.5\phi_d + 10.05\phi_d^2 + 0.0027\exp(16.6\phi_d). \tag{6.5.5}$$

When η_d and η_c are the viscosities of the dispersed and continuous phase, the relative viscosity is given by

$$\eta_r = 2.5\left(\frac{\eta_d + 0.4\eta_c}{\eta_d + \eta_c}\right)\phi_d. \tag{6.5.6}$$

As in the case of foams where $\eta_d \leq \eta_c$ then the term within the parenthesis becomes 0.4 and the relation modifies to $\eta_r = \phi_d$ while in the case where $\eta_d \geq \eta_c$ the term within the parenthesis equals unity and the above relation modifies to $\eta_r =$

$2.5\phi_d$. Other factors that affect the rheology of emulsions are the physico-chemical properties of the continuous phase, its viscosity (η_c), pH, and the film's interfacial properties.

6.6 Instability of Emulsions

A clear and undeniable signature of an unstable emulsion is its loss in homogeneity and phase separation over time into distinct layers irrespective of whether it is an O/W or W/O emulsion. As evident from thermodynamics, $\Delta G > 0$ for emulsions indicates the system is unstable. As a matter of fact, emulsions are unstable and actually separate out over time which may vary from a few minutes to a few months. Four processes namely (1) Creaming, (2) Flocculation, (3) Coalescence and (4) Ostwald ripening contribute toward an emulsion being unstable [14–17]. Figure 6.4 shows the schematic of the four processes that contributes to instability of emulsions.

6.6.1 Creaming and Sedimentation

In an O/W emulsion, creaming involves the rising of oil droplets under gravity to form a concentrated oily layer at the surface. On the other hand, the separation of the water layer at the bottom in a W/O emulsion is called sedimentation. Creaming and sedimentation are reversible processes where the emulsion may be re-obtained with the original uniform droplet distribution.

Considering the droplets to be spheres of radius '*r*', the particle experiences a gravitational force equal to the weight of the droplet that is balanced by the buoyant forces and viscous drag that act in opposite directions. Considering the balance of

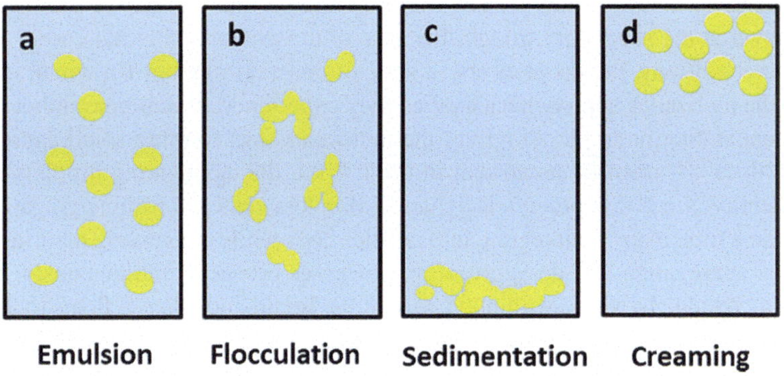

Fig. 6.4 **a** Schematic of a stable emulsion, **b–d** various instable conditions blue represents water, and yellow represents oil

forces, at equilibrium we have $F = \frac{4}{3}\pi r^3(\rho_d - \rho_c)g = 6\pi \eta_0 r v_0$

$$v_0 = \frac{2r^2(\rho_d - \rho_c)g}{9\eta_0}. \tag{6.6.1.1}$$

where v_0 is the resultant terminal velocity of the droplet, r is the radius of the droplet, ρ_d and ρ_c are the bulk densities of the dispersed and continuous phases, respectively. When $\rho_d < \rho_c$, then v_0 is negative which means that the droplets will float up to the top forming a cream and when $\rho_d > \rho_c$ then v_0 is positive which indicates that the droplets will sink to the bottom of the liquid forming sediments. These simple calculations are approximate and holds good for very dilute emulsions corresponding to the phase volume of the droplet medium $\phi_d \approx 0.01$ while for moderately concentrated emulsions $(0.2 < \phi_d < 0.1)$ the velocity of the droplet modifies to $v = v_0(1 - k\phi_d)$ where k is a constant that accounts for hydrodynamic interactions and is usually of the order of 6.5. For highly concentrated emulsions $(\phi_d > 0.2)$ the rate of creaming and sedimentation becomes a complex function of ϕ_d and using the virial expansion $v = v_0\left(1 + a\phi_d + b\phi_d^2\right)$ where the constants a and b are characteristic of that system and may be determined experimentally.

6.6.2 *Flocculation*

Figure 6.5 shows a plot of potential energy $V(r)$ as a function of the distance between the centers of two spherical droplets approaching each other and the only dominating interactions are (a) electrostatic forces of repulsion and (b) attractive van der Waals forces. The net potential is a sum of the individual potentials and is given by the black line while the electrostatic and van der Waals potentials are given by red and blue lines.

Clearly, there are two well-separated troughs, a deep one marked $(V1)$ at $r = r_1$ and known as the primary trough and a shallow one marked $(V2)$ at $r = r_2$ and known as the secondary trough. Because of the available thermal energy under ambient conditions, the droplets are in state of constant random Brownian motion and as the molecules approach each other, they experience attraction as indicated by the potential energy graph. At $r = r_2$ the molecules start forming small aggregates called "flocs". It must be noted that in these flocs, the aggregating droplets retain their identity. Since this process is dynamic, droplets gather to form aggregates and at the same time many disintegrate into smaller flocs while conserving their identity. Both the aggregating and de-aggregating processes proceed simultaneously. These flocs may reside anywhere within the bulk of the liquid and when it floats to the top enhancing creaming. Formation of flocs is a clear indicator of the instability of an emulsion. Mathematically, flocculation obeys a second-order kinetics and is given by the differential equation

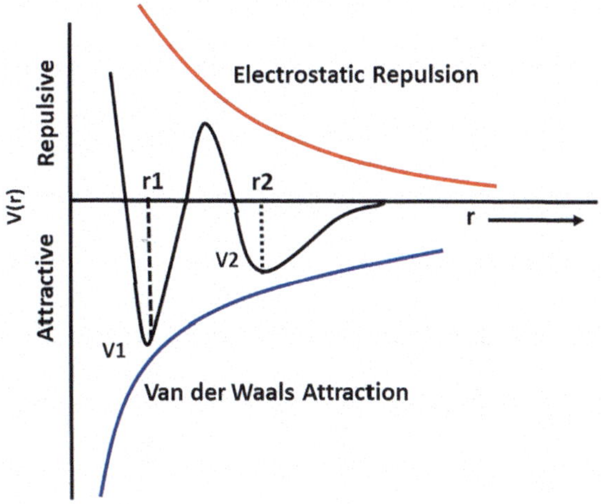

Fig. 6.5 Schematic representing the various interactions stabilizing an emulsion

$$\frac{dn}{dt} = -k_1 n^2 \tag{6.6.2.1}$$

that provides a solution $n = \frac{n_0}{1+k_1 n_0 t}$ where n is the number of flocs at any point in time while n_0 is the number of flocs at $t = 0$, k_1 is the aggregation rate constant, and is given as $k_1 = \frac{4k_B T}{3\eta}$ where k_B is the Boltzmann constant, η is the viscosity and T is the temperature. An important outcome of this calculation is the flocculation half-life $t_{1/2}$ which indicates the time after which 50% of all the particles have formed flocs. Higher the value of $t_{1/2}$ greater the stability of the emulsion against flocculation [11, 17, 18].

6.6.3 Coalescence

This is an irreversible process in which two or more emulsion droplets fuse together to form a single larger entity. The necessary condition that permits coalescence to occur is that the drops must be in contact or in very close proximity to each other at V_1 with $r = r_1$ as shown in Fig. 6.5. Detailed studies have revealed that as the droplets approach each other, flattening of the approaching interfaces occurs and a very thin film of the continuous medium separating the two droplets is created. Presence of thermally generated oscillatory waves in the thin film causes it to rupture and create a hole as shown in Fig. 6.6 allowing liquid from the smaller droplet to get transferred to the larger droplet as the internal pressure in the smaller droplet is higher.

Reflective light scattering has confirmed the presence of oscillatory waves in the thin film. Thus the smaller drop will become smaller over time and ultimately vanish while the larger one grows at the expense of the smaller one and this process continues till one large drop remains that eventually floats to the top as cream or settles down as sediment [14–18]. Alternatively, holes may be formed spontaneously because

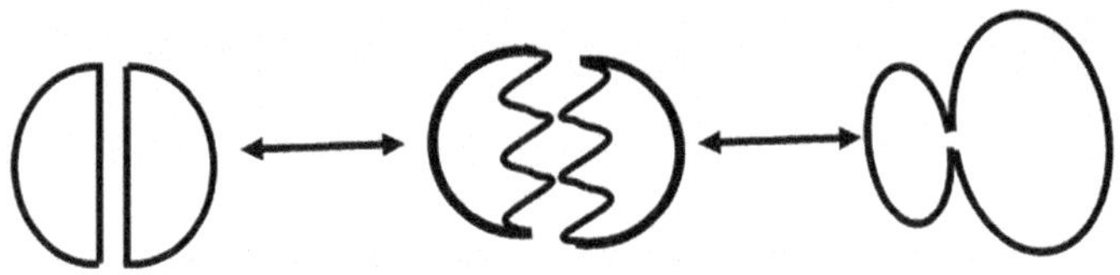

Fig. 6.6 Schematic of coalescence process

of thermal motion in the system. However, when the hole is smaller than some critical value it gets healed but above this critical value the hole increases in size, and film rupture occurs. Lower emulsifier concentration or non-uniform distribution of emulsifier concentration is caused during the high shear emulsification process. In the absence of a surfactant or polymer at the interface of the droplet and continuous medium, the drainage of the fluid between the droplets is fast due to Gibbs–Marangoni effect, and van der Waals forces become intensified that accelerates the coalescence process. Inclusion of a surfactant or a polymer dampens the oscillation causing the rate of coalescence to slow down.

6.6.4 Ostwald Ripening

According to Lord Kelvin, small drops will dissolve and get deposited on larger ones resulting in the disappearance of small droplets and growth of the larger ones leading to coarsening of the emulsion which if uncontrolled will lead to phase separation over time. Needless to mention for a strictly monodisperse system coarsening cannot occur hence it is manifest only for heterogeneous disperse systems.

Mathematically, the bulk solubility of a small droplet of radius r is given by $S(r)$ where

$$S(r) = S(\infty)\exp\left(\frac{2\gamma V_m}{rRT}\right) = S(\infty)\exp\left(\frac{\alpha}{r}\right) \quad (6.6.4.1)$$

where $S(r)$ = Bulk Solubility of the droplet of radius r, $S(\infty)$ = is the solubility of an oil droplet for a planar oil–water system, T = Temp in °K, V_m = Molar Volume, R = Gas constant, γ = Interfacial tension (IFT). The term $\alpha = \frac{2\gamma V_m}{\mathrm{RT}}$ is the characteristic length. The growth rate of a crystal as a result of Ostwald ripening is given by the equation

$$\frac{\mathrm{d}r_c^3}{\mathrm{d}t} = \frac{4}{9}\left[\frac{S(\infty)\alpha D}{\mathrm{RT}}\right]F(\phi) \quad (6.6.4.2)$$

where $F(\phi)$ is a factor that depends on the disperse volume fraction and r_c is the critical radius that is neither growing nor decreasing. Thus droplets with $r > r_c$ would tend to grow into bigger droplets while droplets with $r < r_c$ would tend to dissolve.

Over time coarsening of the emulsion and separation into two distinct layers will occur. The rate of droplet growth may be reduced by lowering the IFT by addition of surface active ingredients like surfactants or polymeric surfactants that will reduce α and the velocity of droplet growth. Adding A-B-A type block copolymers that are soluble in the oil phase but insoluble in the continuous aqueous phase will be readily adsorbed at the oil–water interface that will lower the interfacial tension that will arrest Ostwald's ripening [11, 18].

6.7 Stability of Emulsions

Most emulsion-based food items like mayonnaise, sauces, etc., have limited shelf life that too under stringent storing conditions as laid down by food and drug agency of that country. Hence stability is crucial for any emulsion-based product. Instability arises from the droplets coalescing to form large aggregates leading to phase separation. Therefore, to ensure stability, the droplets must remain apart from each other at all times. Various possibilities exist, one being the electrostatic repulsion between similarly charged droplets that prevents close approach of the droplets and their aggregation. Encapsulated droplets or polymers or particles adsorbed on the droplets are known to provide steric hindrance preventing aggregation. Two opposing forces contribute toward aggregation of particles in an emulsion. While the Van der Waals forces are short range and attractive in nature facilitating close approach of the droplets and their aggregation, the opposing force is the repulsive electrostatic forces. A sum total these interactions finally determines the net scenario of whether or not aggregates will be formed and if the emulsion would be stable.

6.7.1 Electrostatic Stabilization

Electrostatic repulsion between similarly charged droplets provides a powerful technique of creating a barrier to close approach or contact between droplets. A prerequisite to the successful achievement of a highly stable emulsion is that the surfactant or polymer must be well adsorbed on to the droplet in such a way that the charge resides on the surface of the droplet. Short-chain anionic and cationic surfactants or polymeric surfactants are well suited for the purpose as shown in schematic Fig. 6.7a.

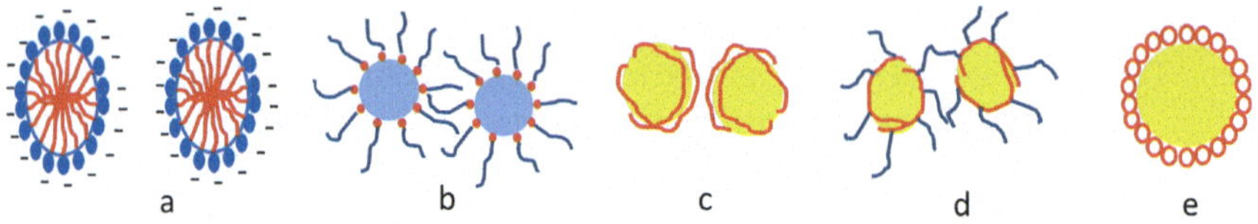

Fig. 6.7 Schematic representing the various interactions stabilizing an emulsion **a** electrostatic repulsion, **b** steric hindrance through polymer adsorption, **c** steric hindrance by long chan in polymers, **d** steric hindrance by polymeric surfactants, **e** Pickering emulsion

6.7.2 Stabilization Through Steric Hindrance Provided by Surfactants and Polymers

Surfactants and Polymers adsorbed on the surface of the droplets provide a mechanical barrier to coalescence as shown in Fig. 6.7b–d. For a O/W emulsion, nonionic surfactants perform well as the polyoxyethylene (POE) long chain remains solubilized while the hydrophobic head groups remain strongly anchored to the oil droplet. Similarly, polymers with a short hydrophobic group and a long hydrophilic chain works excellently. For W/O emulsions the reverse arguments hold well.

6.7.3 Stabilization Through Steric Hindrance Provided by Particles

Since keeping the droplets apart is key to making an emulsion stable, the same may be achieved by lodging inert fine particles at the oil–water interface. Because the interfacial tension between the two liquids is very low, Gibbs free energy (ΔG) is very small at the interface between two liquids and hence small particles tend to get irreversibly adsorbed at the interface as shown in schematic Fig. 6.7e. Theoretically, it has been shown that Gibbs free energy at the interface is given by

$$\Delta G = \pi r^2 \gamma_{12} (1 - \cos \Theta)^2 \tag{6.7.3.1}$$

where r is the radius of the particle, assuming it to be spherical, Θ the contact angle and γ_{12} is the interfacial tension [19–23]. The maximum value of $\Delta G = 2\pi r^2 \gamma_{12}$ suggests that a substantially large amount of energy is required to dislodge a particle from the interface. In fact, nanoparticles adhering to the interface forms a rigid monolayer that provides mechanical barrier to coalescence. Such stabilization by finely divided particles even in the absence of surfactants do stabilize emulsions and are referred to as "Pickering Emulsions". Such highly stable emulsions exist abundantly in nature and find many applications in industry like separation of mineral particles by froth flotation where tiny particles of metal are made to float to the surface by bubbling air through a slurry [19, 20].

6.8 Phase Inversion in Emulsions

As mentioned above, emulsions are either O/W or W/O and under special conditions, the O/W transforms into a W/O emulsion and vice-versa. The phase transition processes are categorized into two different groups (a) Catastrophic and (b) Transitional [11, 17, 18].

6.8.1 Catastrophic Inversion Phase (CIP)

Phase inversion of emulsions may be brought about by changing the dispersed phase (ϕ_d) volume fraction by increasing the volume fraction of oil (ϕ_0) in an O/W emulsion converting it to a W/O emulsion as shown in schematic Fig. 6.8a–e. The specific value ϕ_d at which inversion occurs is referred to as the emulsion inversion point (EIP) may be readily identified through conductivity measurements and is defined as the ratio of volume of aqueous phase at phase transition to volume of the oil phase. Catastrophic inversions are generally irreversible and EIP for an O/W emulsion is different from a W/O emulsion but depends on several factors like agitation speed, rate of addition, temperature, and HLB of the emulsion system, and electrolytes present. In general, EIP decreases with an increase in HLB number and reaches a minimum at which the phase inversion occurs.

Furthermore, at EIP the viscosity is highest and interfacial tension (IFT) lowest [11, 18] as shown in schematic Fig. 6.9.

At high values of $\phi_d = 0.74$, the emulsion is referred to as high internal phase emulsion (HIPE) [24]. A close inspection of Figs. 6.2b and 6.9c clearly demonstrates the difference between an emulsion at low and high values of ϕ_d. High internal phase pickering emulsions (HIPPE) have also been reported [25]

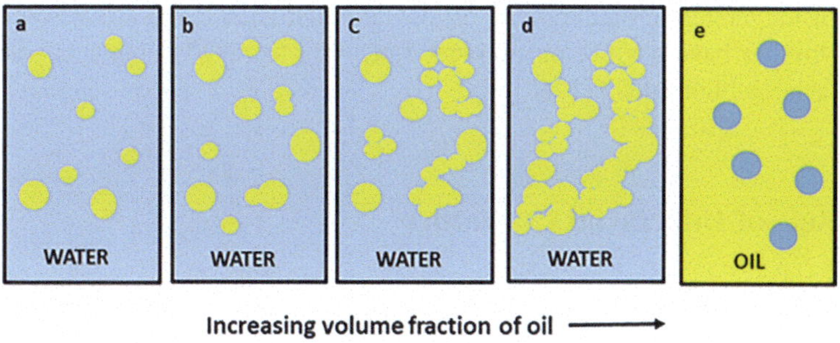

Fig. 6.8 **a–e** Shows the sequences of catastrophic transition from an oil-in-water to a water-in-oil emulsion

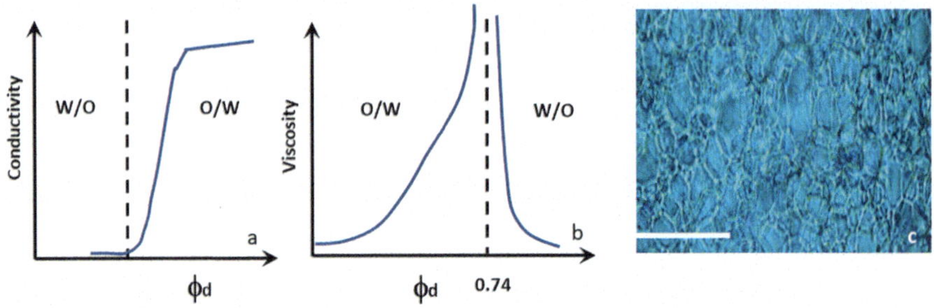

Fig. 6.9 **a** Schematic of conductivity versus oily phase volume fraction (ϕ_d), **b** schematic of viscosity versus oily phase volume fraction (ϕ_d), **c** Microscopic image of the emulsion at $\phi_d = 0.74$. White bar represents 20 μm

6.8.2 Transitional Phase Inversion

Transformation of an O/W emulsion to a W/O emulsion may be brought about by changes to the molecular geometry of surfactant molecules in response to changes in ionic strength, temperature, effective HLB number of the surfactant system, and type of surfactant used.

Nonionic polyoxyethylene (POE) surfactants are special as the POE chains tend to get dehydrated with increasing temperature and its aqueous solubility decreases resulting in "clouding" of the surfactant solution. This temperature at which clouding occurs is called the "cloud point" which is specific to the surfactant under consideration. An O/W emulsion at low temperature when heated becomes unstable and undergoes inversion to a W/O emulsion that can be readily monitored through electrical conductivity measurements and the temperature at which the phase inversion occurs is called the phase inversion temperature (PIT). Ionic surfactants also show similar behavior as above when concentration of the electrolyte is increased.

An alternative way of triggering phase transition is by changing the HLB of the system using an appropriate mixture of low and high HLB components. When the HLB is low, we have a W/O emulsion that transforms to an O/W emulsion when the HLB reaches a high value [11, 18].

6.9 Role of Interfacial Curvature

Phase transition from an O/W to W/O emulsion irrespective of the process catastrophic or transitional goes through the same sequence of change in curvature of the droplets. While a convex interface yields an O/W emulsion, a concave interface yields a W/O emulsion. Thus the curvature of the interface decides whether or not the emulsion will be an O/W or a W/O. It is often convenient to conceptualize the role of curvature using the packing parameter p, where $p = \frac{V}{al}$ and is the ratio of the volume (V) of the hydrophobic tail and the product of the polar head group (a) and

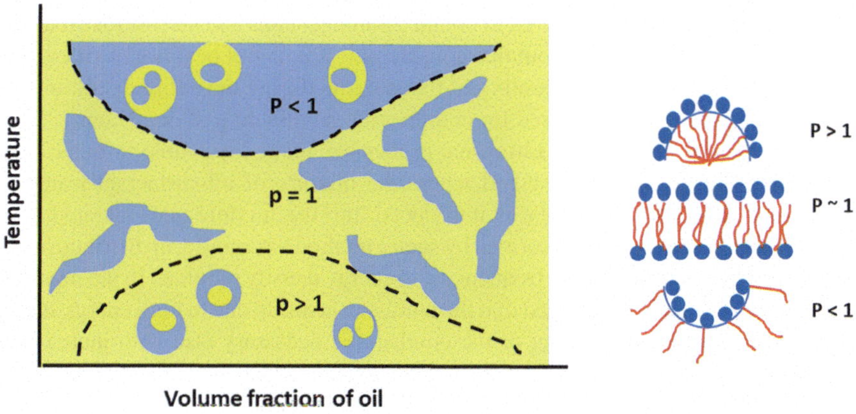

Fig. 6.10 **a** Schematic of various phases at different temperatures and at different oil volume fractions. Phase transition occurs when $p \sim 1$ is through the bicontinuous phase. **b** Molecular arrangement of the amphiphiles at different values of p

(l) is the chain length. While $p > 1$ implies a O/W emulsion, $p < 1$ indicates a W/O emulsion, and $p \sim 1$corresponds to a bilayer. Transition from an O/W to W/O emulsion will obviously involve a transition from $p > 1$ to $p < 1$ or vice-versa through $p \sim 1$ and occurs readily when p is close to 1 as shown in Fig. 6.10.

6.10 General Method for Preparing Emulsions

As evident from thermodynamic considerations, changes to the Gibbs free energy is positive ($\Delta G > 0$) for all emulsions (nano-, mini-, and macro-) suggesting that input of energy is needed to disperse droplets of one phase in the other. An exception to the above are microemulsions where changes to the Gibbs free energy is negative ($\Delta G < 0$) and dispersion of oil in water or vice-versa occurs spontaneously. To make an O/W emulsion a water soluble emulsifier is required while for a W/O emulsion an oil soluble emulsifier is essential. This fundamental above requirement is the underlying statement of Bancroft's rule. A typical surfactant concentration between 5 and 10% w/w is usually required and depending on the type of emulsion needed, a suitable emulsifier is chosen. For a W/O emulsion, a low HLB emulsifier (HLB 1–10) is selected and completely dissolved in the oil phase under constant low speed (20–50 rpm) stirring. Then a pre-calculated amount of water determined from the O/W ratio is slowly added to the oil under constant mixing speed mixing of about 300–1500 rpm till a uniform and homogeneous mix is obtained. A similar procedure is followed for the O/W preparation except that the emulsifier should have HLB in the range 30–40. Emulsions with low polydispersity is often required in cosmetics and pharmaceutical products that is achieved by using a high speed (6–15 K) rpm rotor–stator system that splits the dispersed droplets into a fine monodisperse emulsion.

Several passes are usually required to obtain a homogeneous and narrow distribution of droplets. It must be pointed out that considerable amount of heat is generated due to frictional forces and hence cooling of the material during the entire operations is a must. This process is often used for commercial manufacture of emulsions.

Availability of high-energy ultrasonic probes has made it possible to make fine emulsions. This is a simple method where the mixture of oil-surfactant-water is placed in a vessel and the probe is introduced into the mixture and operated at a given frequency. The emulsion formed by such a method has a uniform distribution of droplet size that depends on the frequency and energy density provided to the mixture by the probe. It must be cautioned that the emulsion heats up due to high energy input by the probe and hence cooling of the emulsion is necessary else with increasing temperature enhanced Brownian motion accelerates growth of large droplets that will eventually separate out as distinct layers of oil and water.

References

1. Tadros TF (ed) (2013) Encyclopedia of colloid and interface science. Springer-Verlag, Berlin
2. Vincent B (2010) Emulsions (chapter 6), 2nd edn. In: Cosgrove T (ed) Colloid science principles, methods and applications. Wiley, U.K.
3. Clausse D (1983) Dielectric properties of emulsions and related systems in Encyclopedia of emulsion technology, vol 1. In: Becher P (ed). Marcel Dekker, N.Y.
4. Erbil HY (2006) Surface chemistry of solid and liquid interfaces. Blackwell Publishing, UK
5. Stauffer D, Aharony A (1992) Introduction to percolation theory. Taylor and Francis, London
6. Sahimi M (1994) Applications of percolation theory. Taylor and Francis, London
7. Panda M (2023) Percolation, scaling, and relaxation in polymer dielectrics. Springer, Switzerland
8. Brinkman HV (1952) J Chem Phys 20:571–592
9. Roscoe R (1952) Br J Appl Phys 3:267–269
10. Kreiger IM, Dougherty TJ (1959) Trans Rheo Soc 3:137
11. Tadros TF (2016) Emulsions: formation, stability, industrial applications. De Gruyter, Germany
12. Ball RC, Richmond P (1980) Phys Chem Liquids 9:99
13. Thomas DG (1965) J Coll Sci 20:267
14. Dukhin SS, Sjoblom J, Saether O (2006). In: Sjoblom J (eds) Emulsions and emulsion stability, 2nd edn. CRC Press, Boca-Raton, Florida
15. Eastman J (2010). Emulsions (chapter 3), 2nd edn. In: Cosgrove T (ed) Colloid science principles, methods and applications. Wiley, U.K.
16. Verwey EJ, Overbeek JTG (1948) Theory of stability of lyophobic colloids. Elsevier, Netherlands
17. Holmberg K, Jonsson B, Kronberg B, Lindman B (2003) Surfactants and polymers in aqueous solution, 2nd edn. Wiley. N.J.
18. Clements DJ (2016) Food emulsions, principles, practice and techniques, 3rd edn. CRC Press, N.J.
19. Robin B, Agnely F, Tsapis, N, Huang N (2022) Chapter 3 in developments in clay science, clay minerals and synthetic analogues as emulsifiers of Pickering emulsions, vol 10. In: Wypych F, de Freitas RA (eds). Elsevier, Netherlands
20. Binks BP, Horozov TS (2006) Colloidal Particles at liquid Interfaces. CUP, N.Y.
21. Binks BP (ed) (1998) Modern aspects of emulsion science. Royal Society of Chemistry, UK
22. He X, Lu Q (2024) Adv Coll Interface Sci 324:103086

23. Abdullah, Weiss J, Ahmad T, Zhang C, Zhang H (2020) Trends in food sciences and tech. 106, 91
24. Yu J, Zhang Y, Zhang R, Gao Y, Mau L (2024) Int J Biol Macromolecules 254:127815
25. He A, Liu Q (2024) Adv Coll Interface Sci 324:103086

Chapter 7
Hydrogels

7.1 Introduction

Webster twentieth century dictionary defines a "Gel" as a "jellylike" substance formed by a colloidal solution in its solid phase, wherein "jelly" has been described as a soft, resilient, partially transparent semi-solid gelatinous food resulting from the cooling of fruit juices boiled with sugar or meat juice cooked down. Although the above gives a convincible description of what a gel is like, it seems incomplete given the complexity of its structure and physico-chemical properties as we know it today. The terminology "Gel" originated from the Greek word "Gelu" which conveys the idea of a liquid setting into a semi-solid that does not flow freely. Gels are readily found in nature and have been known to man since ancient times through centuries. It is found in plant and animal tissues. These entities hold large volumes of biological fluids rich in nutrients, anti-oxidation agents, and electrolytes that support the living system. In the human body, the cornea, vitreous humor, connective tissues, and inner lining of our stomach, lungs, kidneys, and blood vessels contain gels that prevent toxins from entering our vital organs but readily transport water, ions, and nutrients to these organs. Interestingly, in the late 1950s use of copolymers of 2 hydroxyethyl methacrylate (HEMA) and ethylene dimethacrylate (EDMA) was used in soft contact lenses [1] triggered and re-ignited interest in a better understanding of gels and their functions.

Various researchers have described "Gels" quite differently each having its own set of merits and discrepency as well. Herman [2] described "gels" as a two-component dispersed system that exhibits mechanical properties that are characteristic of a solid and where both the dispersed component and the dispersion medium extend throughout the whole system. Amdal [3] defined "gel" as a two- or multicomponent soft solid where at least one is a liquid. In addition, the storage modulus (G') with time shows a plateau while the loss modulus (G'') is much lower in value in the same plateau region. Perhaps the simplest and an outstanding description has been

A. K. Dutta, *Soft Matter*, https://doi.org/10.1007/978-981-96-0624-5_7

provided by Nijenhuis and Nishinari [4] who states "A gel is a gel, as long as one cannot prove it otherwise".

Gels consist of at least two ingredients, usually a polymer that forms a 3D network and a solvent that remains trapped in the network and may vary from 10 to 1000 times the weight of the polymer. The polymeric network forming the scaffold and the liquid entrapped contributes to its mechanical properties like softness and ability to deform under externally applied forces. When the entrapped liquid is water, we have a "Hydrogel". On replacing the entrapped liquid with a hydrocarbon solvent or oil we get an "Organogel" or "Oleogel". Hybrid systems where a mix of water and a hydrocarbon is used yield "Bigels". Similarly, when the scaffold is filled with a gas or air we have an "Aerogel".

Recent years have seen a surge in interest in gels both from academia and industrial research because of its versatile and diverse applications that range from dental cavity fillers to controlled and targeted drug release that is invaluable in the treatment of many diseases. Because hydrogels hold considerable amounts of water as in human tissues, medicated hydrogels have been used for burns and wound treatment. Smart or intelligent gels that respond to external stimuli like pH, temperature, ionic strength, and solvent mixtures have been used in the design of biosensors and smart drug delivery systems. In this chapter, we briefly explore the physico-chemical properties of hydrogels and their various applications. Smart and Intelligent gels and their applications is biomedical devices and in electronic and photonic devices have also been briefly described.

7.2 Types of Gels

Based on the descriptions above, various types of gels are listed below.

(a) **Aerogels**

These are materials synthesized by the sol–gel route. Instead of a solvent that fills the pores of the gel, a gases or air is used. Aerogels may have many interesting features like densities varying from 1 to 1000 kg/m^3, high specific surface area, refractive index, dielectric constant, and ultralow conductivity to mention a few. Aerogels have been prepared from synthetic polymers like polyvinyl alcohol, and polyamides, or natural polymers like cellulose, resins, and polysaccharides. Depending on the condition of their preparation, these systems may represent colloidal nanoparticle, nano- or micro-fibrillar systems [5].

(b) **Bigels**

These consist of two immiscible and independently formed gels that have been stabilized. It is equivalent to an oleogel and a hydrogel fused together which makes these systems unique in that incorporation and release of both hydrophilic and hydrophobic drugs is possible [6]. These features make designing of sequential delivery of multi-functional actives as in cosmetics and drug delivery possible. Moreover, manufacturing bigels is simpler and cost-effective compared to nanoparticles or nano-liposomes for the same purposes and hence is a preferred format for cosmetics and medicine.

(c) **Cryogels**

These are formed by polymer precursors reacting at sub-zero temperatures yielding a highly interconnected macroporous gel [7, 8]. Its formation occurs sequentially as follows: (a) phase separation of the organic phase with ice crystals formed at that temperature, (b) crosslinking and polymerization and (cS) thawing of the ice crystals. These steps form a mechanically stable, highly porous and interconnected scaffold that finds many biomedical applications.

(d) **Emulgels**

Emulgels as the name implies are emulsion-based gels and may either be an oil-in-water (O/W) or a water-in-oil (W/O) type and are prepared by adding a gelling agent to an emulsion. Such systems have been used for controlled drug release in medicine and cosmetics. The internal phase of the emulsion droplets acts as reservoir for the actives that get released through a diffusion-limited process. The rate of release of drugs from the internal phase depends on the composition of the emulsion, the hydrophobicity, hydrophilicity of the drug, viscosity of the gel, and temperature. Because of their thixotropic characteristics, spreading ability, ease of extrusion, grease-free emollient, and long shelf life these are preferred over conventional emulsions [9, 10].

(e) **Hydrogels**

Hydrogels are formed when a gelator, above a threshold concentration is dissolved in an aqueous or polar or its mixture is used as a solvent. The gelator molecules then self-assemble to form a 3D porous network in which the liquid is immobilized. Various interactions like van der Waals, H-bonding, and electrostatic contribute toward the formation of self-assembled structures that form the backbone of the gel structure [11–13].

(f) **Organogels**

These are identical to hydrogels except that the continuous liquid phase immobilized in the 3D network is organic [14, 15]. The organogelators that form the polymeric network consist of lecithin, glyceryl fatty acid esters, and sorbitan mono-stearate which are the most common. These are suitable as delivery vehicles for essential oils, fatty acids, and hydrophobic drugs for transdermal applications as these systems are excellent dermal permeation facilitators.

(g) **Xerogels**

These are prepared by reacting a metal alkoxide precursor with water and alcohol. On slow drying at ambient conditions, the solid material formed has very small pores, large surface area, and high porosity that make these materials suitable as separating membranes. Moreover, these materials are known to be safe, biocompatible biodegradable, and used in various tissue engineering applications [16, 17].

7.3 Classification of Gels

There are several ways of classifying polymer gels, a convenient way is to categorize them into two main groups: (a) Physical Gels and (b) Chemical Gels based on the nature of interactions with the components in the gels.

(a) Physical gels and their stabilization in solvents are achieved through weak forces namely hydrogen bonding, van der Waals and hydrophobic-hydrophobic interactions among others. The 3D networks are also formed through entanglements of polymeric chains through twisting and braiding with each other and forming knots as shown in schematic in Fig. 7.1a–d [6, 7].

(b) Chemical gels are formed as a result of covalent crosslinking of various polymeric segments that establishes a 3D network as shown in schematic. Addition polymerization of oligomeric multifunctional precursors, vulcanization of high molecular weight linear polymeric chains, end linking of reactive chains with branching units are some of the many mechanisms of producing chemical gels. Crosslinking polymerization as in polyacrylamide and addition or condensation polymerization of silicone derivatives are examples of gel formation by chemical means. Crosslinking may be also achieved thermally or though irradiation with ultraviolet, x-ray or gamma rays.

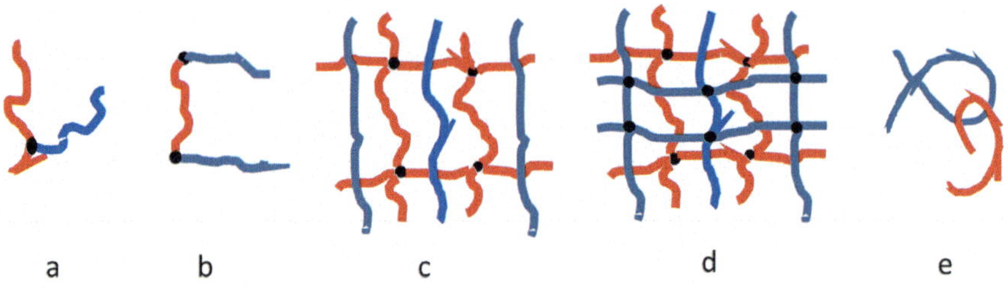

Fig. 7.1 Schematic of **a** copolymer, **b** block polymer, **c** semi-IPN, **d** IPN, **e** entanglement. The two colors correspond to two different polymers

7.4 Gel Preparation

Low molecular weight gelators (LMWG) are used to form gels that are usually prepared by solubilizing the LMWG completely in a suitable solvent or by heating it in the solvent and then allowing the hot solution to cool. The gel formed depends on the LMWG, its concentration, and the solvent used. It is important to point out that gel properties, namely morphology and stability over time depend on the rate of cooling. In fact, substantially different properties have been observed for the same gel when cooled very slowly (1 °C/min) versus rapid cooling by dipping in water at (0 °C). In many cases, gels may be directly formed under ambient conditions without the need to heat or cool. Because gels are metastable, their configuration changes over time and so do the properties on standing. Hence these systems are in state of dynamics equilibrium. While in some cases gels initially formed break down into small aggregates over time, there are others where consistency and physical properties of gels have improved.

7.5 Mechanism of Gel Formation

Polymers in their dry state exist as small coils that in a suitable "good" solvent open up as individual strands. At very low concentrations these chains remain randomly oriented in the solvent matrix. With increasing concentration, the chains come in close proximity with each and the network starts forming either through chemical interactions (covalent bonding) or through physical interactions (H-bonding, van der Waals forces, or simple entanglement of the chains) to form a 3D fibrillary network entrapping a large volume of the solvent that constitutes the gel. The mechanism of gel formation is illustrated in the schematic Fig. 7.2.

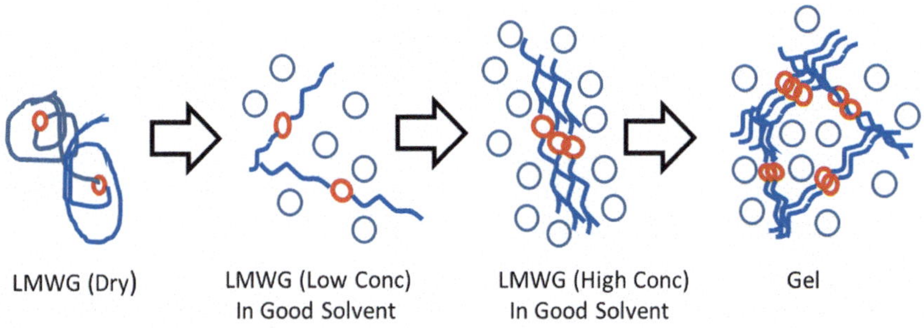

Fig. 7.2 Schematic of gel formation by an LMWG in a Good Solvent. Wavy blue lines correspond to the polymers and red circles are the chromophores. Open blue circles represent Good Solvent

7.6 Physical Characteristics of Gels

7.6.1 Swelling

One of the most distinguishing factors that distinguishes gels from highly concentrated solutions is its ability to swell and hold large amounts of solvent to a tune of 1000 times the weight of the polymer itself. The rate of absorption of the solvent is a slow process that depends on both the polymer and the solvent. Hence the weight of solvent absorbed by a polymer is a function of time and is expressed as $W(t)_{\text{wet}}$ where $t < t_{\text{equilibrium}}$. At $t_{\text{equilibrium}}$ the absorption process is completed and no further increase in weight is observed and W_{Dry} corresponds to the weight of the dry polymer before it comes in contact with the solvent. The swelling ratio (SR) is defined by the relation below:

$$\text{SR} = \frac{W(t)_{\text{wet}} - W_{\text{dry}}}{W_{\text{dry}}}. \tag{7.6.1.1}$$

7.6.2 Equilibrium Water Content

A water-insoluble polymer placed in water absorbs water and swells. The rate of water intake is dependent on the polymer structure and molecular weight. After long enough time, usually taken as 24 h by which time the system is supposed to have reached equilibrium. The wet polymer is carefully removed and the surface water adhering to the polymer is removed by gently touching the surface of the polymer with a tissue or filter paper. Weighing the polymer loaded with water yields W_{e} and the equilibrium water content (EWC) is given by

$$\text{EWC} = \frac{W_e - W_d}{W_e}. \tag{7.6.2.1}$$

7.6.3 Rheological Properties

Rheology provides a measure of resistance offered by a material against deformation. Gels display features typical of viscoelastic materials, in other words, exhibit elastic properties that are typical of a solid and viscous properties that are exhibited by liquids. Therefore, when an oscillatory stress (σ) is applied to a viscoelastic material, the elastic or storage modulus (G') that is characteristic of its elastic property and the loss modulus (G'') that is a measure of its viscous properties [18, 19]. For a

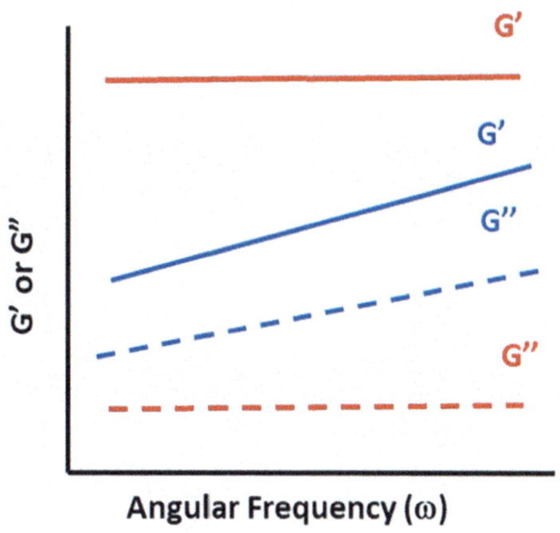

Fig. 7.3 Schematic of G' and G'' for a strong or stiff gel (red) and a weak gel (blue)

viscoelastic material, there exists a phase difference between the applied sinusoidal stress and the corresponding strain. The strain lags in phase by δ where $\text{Tan}\delta = \frac{G'''}{G'}$. When the gel network is such that the bonds are static and permanent as in covalent bonds, there is no dependence of G' and G'' on the angular frequency (ω) while for a dynamic network as in the case of weak and non-permanent bonds (van der Waals, hydrogen bonds) there exist strong dependence on the angular frequency [18, 19]. In hard gels $G' \gg G''$ at all frequencies while in soft gels $G' > G''$ at high frequencies and $G' < G''$ at low frequencies as have been observed for worm like micelles.

The dependence of G' and G'' as a function of time indicates dynamic network in the gels. For a pure elastic solid, G'' is zero, and for a pure liquid G' is zero. Thus when the ratio $\text{Tan}\delta = \frac{G\prime\prime}{G\prime}$ is less than 1, then the gel is strong and when greater than 1, the gel is weak as shown in schematic Fig. 7.3. Larger the gap between G' and G'', stiffer is the gel, and smaller the gap between G' and G'' weaker is the gel.

When the ratio of G''/G' given as $\text{Tan}\delta \leq 0.1$ there is no dependence on the angular frequency and G' and G'' curves are parallel to the angular frequency axis with G' typically 1–2 orders of magnitude higher than G'. However, for $\text{Tan}\delta \geq 0.1$ slight dependence of G' an G'' on angular frequency has been observed for several molecules namely collagen, hyaluronic acid and soft tissues.

7.6.4 Free and Bound Water

When a gel is left to stand, a small volume of the solvent is observed to collect at the surface of the gel and is actually the interstitial liquid expelled (Fig. 7.4) as a result of relaxation of elastic stresses during setting of the gels during which the volume decreases and forces out the interstitial liquid that accumulates at the top of the gel. A reliable way of estimating the % of free water in the sample is to carefully weigh

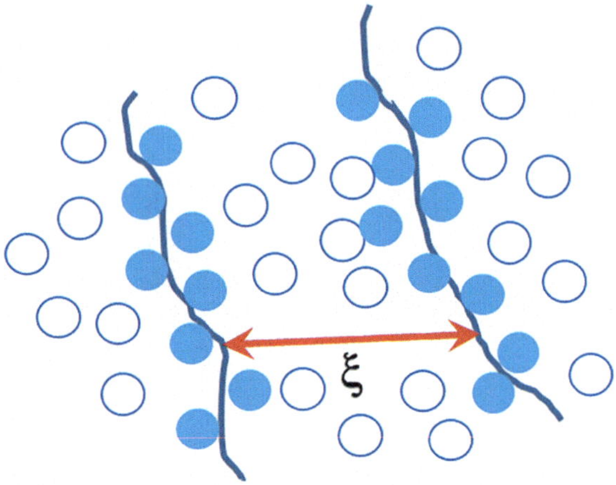

Fig. 7.4 Bound and free water in a gel. The open circles indicate free water and solid circles indicate bound water. ξ is the polymer's average mesh size

out a sample of the gel and centrifuging it at 1500 rpm for about 15 min and then aspirate out the separated water. The difference in weight of the gel gives a fairly accurate estimate of free water in the gel.

7.7 Synthetic Polymer Hydrogels

Figure 7.5 shows molecular structures of some synthetic polymers used as gels for various purposes discussed below.

7.7.1 Polyvinyl Alcohol (PVA)

Polyvinylalcohol (PVA) molecular structure which is shown in Fig. 7.5 is perhaps one of the most extensively used synthetic polymers in the pharmaceuticals and cosmetics industry because of its low oral toxicity, biocompatibility, and bio-degradability profiles. Moreover, it is water soluble with excellent film forming properties [20]. It is however sparingly soluble in other polar solvents like ethanol. Its solubility and degree of hydrolysis depends on its molecular weight. With increasing molecular

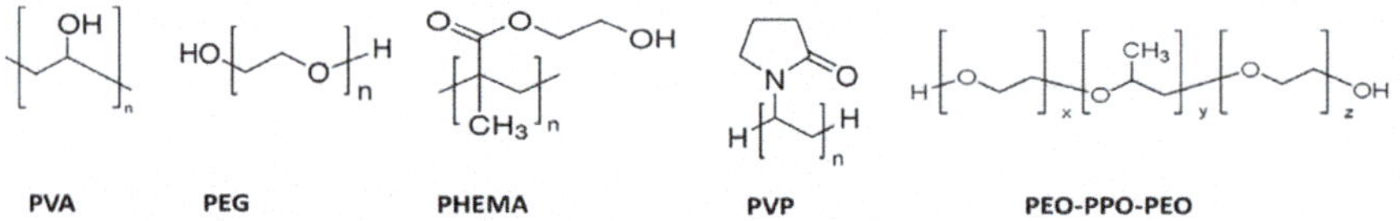

Fig. 7.5 Molecular structures of some synthetic polymers

weight, its crystallinity also increases. Because PVA readily crosslinks via a freeze–thaw process to form a cryogel (PVA-C) it is a polymer of choice in the pharmaceutical industry as crosslinking agents and residues can be safely avoided. PVA is known to react with acrylate copolymers making it amphyolytic through inclusion of both cationic and anionic groups. PVA in the presence of borate ions form weak gels that do not retain their shape. PVA-chitosan–gelatin hydrogels irradiated with gamma radiation have been used in wound dressings and artificial grafts [21].

7.7.2 Polyethylene Glycol (PEG)

Polyethylene Glycol (PEG) is an extensively used antifreeze agent and a diluent in the paint and pharmaceutical industry. PEGs are hydrophilic and readily soluble in water but insoluble in some solvents like ethanol, methanol, and methylene chloride and exhibit amphiphilic properties that allow them to partition into cell membranes. While low molecular weights PEGs are free-flowing liquids, with increasing molecular weight, it becomes viscous, and above 10KDa are soft solids. Like most ethoxylated compounds, PEGs exhibit inverse solubility, i.e., with increasing temperature its solubility decreases, and above 100^{0}C PEG phase separates out. Its low toxicity and immunogenic reactivity, and high biocompatibility make it an excellent ingredient for pharmaceutical applications [22]. PEGylation of drug and protein molecules is known to improve pharmacokinetic and pharmacodynamic responses. Transplantation of living cells and delivering it to target sites are often achieved by encapsulation of cells in PEG diacrylate microspheres. These above-mentioned features make PEG's a preferred material for tissue culture and organ storage [23].

7.7.3 Poly(2-hydroxyethylmethaacrylate) (PHEMA)

PHEMA belongs to polyacrylic acid family and has many important features characteristic of acrylates namely its mechanical and structural strength and transparent appearance that make these excellent materials for contact lenses. Like other acrylates, PHEMA forms films that allow diffusion of gases and liquids that qualify these materials for breathable soft contact lenses and artificial corneal grafts [24]. PHEMA hydrogels are also used for controlled drug and protein release and has been used in the development of artificial skin and breast implants. Because of their biocompatibility, P[illegible] film coatings has been successfully used for various stents and mic[illegible]

7.7.4 Polyvinylpyrrolidone (PVP)

Polyvinylpyrrolidone (PVP) is a polymer that finds use across industries. It is nonionic, water soluble, stable across the entire pH range. It is nontoxic and has excellent film forming properties in addition to being biocompatible that makes it a valued material for biomedical applications. Chitosan-PVP hydrogels has been used for activation and proliferation of endothelial cells in vascular applications and tissue engineering [20, 25].

7.7.5 Pluronics

"Pluronics" is the trade name of $(PEO)_n$-$(PPO)_m$-$(PEO)_n$ block copolymers with the central hydrophobic component PPO flanked by two hydrophilic elements PEO and subscripts m and n are integers. These copolymers are characterized by their low critical micellar concentrations (CMC) usually less than 1% w/w and form gels with different degrees of stiffness depending on the polymer concentration used [26, 27]. The central part of the molecule cluster together to form the hydrophobic core of a micelle while the hydrophilic chains form the periphery and remain in contact with the aqueous phase. Details studies have revealed that within the hydrophobic core, the PPO group remains stretched from one end to the other, and hence with changes in the PPO segment the micellar size changes [27–29]. Again because of the hydrophobic characteristic of these micellar cores these act as excellent solubilizing agents for hydrophobic molecules. A unique feature of these triblock copolymers is that at low concentrations and temperatures below 25 °C, these molecules exist as single strands (unimers) oriented randomly but at higher concentrations these form organized aggregates or micelles but at higher temperatures form gels. It was observed that (a) below polymer weight of 2000 gm/mole no ordered phase exists. (b) Sequence of mesophases depends on the PO/EO ratio. When PO/EO ratio exceeds 0.5, spherical micelles are formed at concentrations greater than the CMC but at very high concentrations, order–disorder transformations occur resulting in the formation of cubic phases. However, at PO/EO ratio equal to 0.25 hexagonal phases are formed while at PO/EO ratio equal to 0.15, lamellar phases are formed. PF-127 (PEO_n-PPO_m-PEO_n) where $n = 100$ and $m = 65$ has been successfully used in various biological applications [30].

7.8 Natural Polymer Hydrogels

Figure 7.6 shows the molecular structures of some common natural polymers used as gels.

7.8.1 *Hyaluronic Acid*

Hyaluronan or hyaluronic acid (HA) was identified by Meyer and Palmer in 1936 in the bovine vitreous humor. Because of its presence in mammalian species, being nontoxic, biodegradable and biocompatible it forms excellent scaffolds for regenerative medicine, drug delivery and cosmetics. HA is a linear glycosoaminoglycan composed of repeating disaccharide units of D-glucuronic acid and N-acetylglucosoamine linked through β-1 $\rightarrow$ 4 and β-1 $\rightarrow$ 3 glucosidic bonds. Methacrylate—Hyaluronic acid photo cross-linked soft hydrogels has been successfully used in for the treatment of cutaneous and corneal wounds [31, 32].

7.8.2 *Chitosan*

Chitin is a water-insoluble natural polysaccharide that is abundantly present in nature, especially in the cell walls of fungi and exoskeletons of crustaceans and insects. N-deacetylated derivative of chitin is chitosan which is however water soluble. Structurally, it is composed of beta (1–4)-2 amino e deoxy-D glucopyranose and beta (1–4) 2-acetamido-2-deoxy-D-glucopyranose units randomly distributed along the polymer chain. Chitosan has mucoadhesive properties and is used in various biological applications [33, 34].

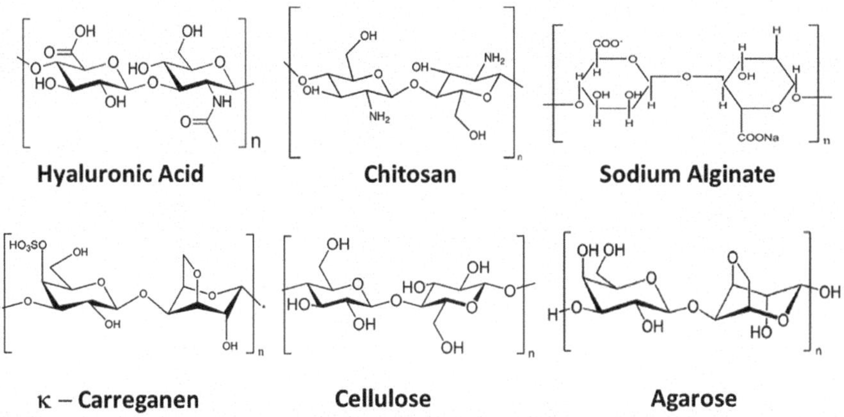

Fig. 7.6 Molecular structures of some natural polymers

7.8.3 Alginates

These are natural anionic biopolymers found in sea weeds and brown algae (Phaeophceae). Its non-toxicity, biocompatibility, and ability to form hydrogels make these important biomaterials that find immense use in foods, beverages, biotechnology, and regenerative medicine. Alginates are block copolymers consisting of alternating mannuronic (M-) and guluronic (G-) acid blocks the ratio of which depends on its source, season of growth and harvest, and the part of the plant used for extraction. All alginates are polydisperse with respect to their molecular weight and the polydispersity index (PI) is given by the ratio of M_w to M_N, where M_w corresponds to the average molecular weight and M_N the number average molecular weight. M_w is usually about 200 K Da and may reach values of 400 K–500 K Da. In presence of divalent cations (Ca^{2+}, Ba^{2+}, Sr^{2+}, Mg^{2+}) alginates at 1–2% w/w in water readily form gels. The driving force behind the formation of gels is the binding of the divalent charges to two chains of the alginate. The binding affinity is given as $Mg^{2+} < Mn^{2+} < Ca^{2+} < Sr^{2+} < Ba^{2+} < Cu^{2+} < Pb^{2+}$. Moreover, that retain large volumes of water or biological fluids that make them excellent matrices for drug delivery systems or entrapment of biomolecules namely proteins or cells [35, 36].

7.8.4 Carrageenan

These are sulphated polysaccharides obtained from red seaweeds (Rhodophyceae) and consist of repeating galactose units and 3,6 anhydrous galactose both sulphated and non-sulphated and alternating units of alpha 1–3 and beta 1–4 glycosidic linkage three varieties are known and identified as Kappa (κ), lambda (λ) and Iota (ι) Lambda Carrageenan does not form gels at all. κ-Carrageenan gels are known to be opaque and brittle in presence of potassium ions. However, ι-Carrageenan gels are transparent soft, and elastic. κ-Carrageenan finds extensive use in pharmaceuticals industry. Formation of IPN beads of κ-carrageenan and sodium carboxy methyl cellulose with $AlCl_3$ as crosslinking agent has been reported. κ-carrageenan with sodium alginate where Ca^{2+} ions were the crosslinking agent has been used in various applications [37, 38].

7.8.5 Aragose

Aragose is obtained from red sea weeds (Gelidium and Gracilaria species). These consist of alternating 1,3 linked galactopyranose and 1,4 linked 3,6 anhydro–alpha-galatopyranose units similar to Carragennans. These It dissolve in boiling water and form gels on cooling it to 35–40 °C. It forms a random coil configuration at elevated temperatures but on cooling attains a helical structure forming a strong gel [39, 40].

7.8.6 *Cellulose*

Cellulose is a hydrophobic polymer that is widely found in plants and in some bacterial species (Acetobacter xylinium). Esterification or etherification can be readily converted into derivatives the hydrophilicity and hydrophobicity of which can be modulated as required. Because of its nontoxic, high biocompatibility, and biodegradability, it is extensively used in biomedical applications. It has been successfully used in cell transplantation, promotion of angiogenesis, treatment of burns, and cellular encapsulations [41, 42].

7.9 Smart Hydrogels

Extensive research on hydrogels over the past three decades has not only helped a better understanding of the structure–property correlation in gels but has also significantly contributed toward advancement of technologies and development of devices like those described below. Smart gels are referred to those that exhibit properties that may be utilized to perform some specific function in response to external stimuli like exposure to stress or strain, electric or magnetic field or light.

7.9.1 *Energy Storage in Hydrogel Batteries*

Polymer gels are known to have a porous matrix that is filled with a solvent, water in the case of hydrogels. Recently, PEO-PPO-PEO polymers have been used with the advantage that with increasing temperature they undergo a phase transformation into a gel state that hinders ionic conduction and effectively switches off the device preventing explosions or damage to devices it is connected to thus making the battery safe to use [43].

7.9.2 *Hydrogel-Based Motion Sensors*

Wang et al. [44] designed a polyvinyl alcohol (PVA) based hydrogel containing maleic acid (MA) and calcium chloride. Such a combination is operated as a small battery that can deliver 0.55 V. Therefore, when a strain is applied to the hydrogel the output voltage changes allowing it to be used as a strain gauge.

7.9.3 Temperature Responsive Hydrogels

Since its discovery in 1978 by Prof. Toyoichi Tanaka thermal-responsive hydrogels have been extensively used in tissue engineering. Hydrogels with transition temperatures close to the human body temperature are of special interest for medical applications. Such hydrogels may be divided into two classes (a) positive and (b) negative thermoresponsive hydrogels. The positive thermoresponsive gels swell above a specific temperature (UCST) while the negative thermoresponsive swells below (LCST). UCST and LCST are the upper and lower critical solution temperatures. Poly(N-isopropylyacrylamide) (PNIPAM) has a LCST of about 32 °C while Poly(vinyl caprolactam) (PVCL) has LCST of 31 °C similar to Poly(N-isopropylyacrylamide [45]. However, both have high glass transition temperatures (T_g) of about 140–150 °C and significant thermal hysteresis. Alternatives to these polymers are Poly(oligo ethylene glycol methyl acrylates) (POEGMA) which consists of a polymethyl acrylate backbone with oligo ethylene glycol side chains and Poly(vinyl methyl ether) (PVME) that has two LCST's at 34 °C and 28 °C, respectively [46]. Both POEGMA and PVME are extensively used in biomedical devices.

7.9.4 PH-Responsive Hydrogels

pH-sensitive hydrogels are usually polyelectrolytes that contain acidic or basic groups. They are either anionic, cationic or zwitterionic. Polyacrylic acid (PAA) and poly(methacrylic acid) (PMAA) are anionic and respond to pH, light, and temperature. PAA cross-linked with glycerol or poly(vinyl alcohol) (PVA) via esterification results in the conversion of chemical energy into mechanical work that is manifested through changes in its volume that is reversible and dependent on the pH and degree of crosslinking [47].

7.9.5 Fluorescent Hydrogels

Fluorescence from organic or inorganic molecules provides a powerful tool for detecting ions like Cd^{++}, Ni^{++}, Hg^{++}, and many others that are toxic to aquatic life and environment. Usually, a gelator containing a polyaromatic chromophore like, naphthalene, anthracene, pyrene, and perylene are used to form a hydrogel that is illuminated by an appropriate wavelength, and the emission from the gelator is recorded. When the ions as mentioned above come in contact with the gelator, the fluorescence emission quenches demonstrate the presence of quencher ions. Since the quenching is proportional to the concentration of the ions, the concentration of the ions in the sample may be readily evaluated. Similarly, several anthracene and pyrene derivatives have been used as gelators to detect changes in pH, temperature, and mechanical stress or strain [48].

7.9.6 *Electroactive Hydrogels*

Hamlen [49] first reported the changes in dimensions of PAA-PVA gels containing finely divided platinum particles. Tanaka had examined phase transitions in various gels induced by electric fields. Osada [50, 51] demonstrated that under an electric field, polyelectrolytes undergo mechanical deformation as the mobile ions migrate to the opposite electrode while the polyelectrolyte backbone remains stationary. They successfully developed a gel looper, a polymer gel-based actuator.

7.9.7 *Magnetic Hydrogels*

Ferrogels [52, 53] are known to be responsive to magnetic fields. These are hydrogels of PVA containing microparticles of Fe_3O_4 having a diameter 10–12 μm exhibit unidirectional Hookean magnetoelastic behavior. k-carrageenan doped with barium ferrite shows giant storage modulus reduction before and after application of a magnetic field.

References

1. Nichterle O, Lim D (1960) Nature 185:117–118
2. Herman PH (1949) Gels in colloid science II edited by Kruyt HR 483–651. Elsevier, Amsterdam
3. Almdal K, Dyre J, Hvit S, Kramer O Polym. Gels Netw 1:5–17
4. Nijenhuis KT (1997) Thermoreversible networks. Adv Polym Sci 130, 1
5. Du A, Zhou B, Zhang ZH, Shen J (2013) Materials 6:941–968
6. Cates ME, Clegg PS (2008) Soft Matter 4:2132–2138
7. Lozinsky VL, Galaev IY, Plieva FM, Savina LN, Jungvid H, Mattiasson B (2003) Trends Biotechnol 21:445–451
8. Henderson TM, Ladewig K, Haylock DN, McLean KM, O'Connor AJ (2013) J Mater Chem B 1:2682–2695
9. Asija R (2013) J. Biomed Pharmaceutical Res 2:91–94
10. Yu J, Zhang Y, Zhang R, Gao Y, Mau L (2024) Int J Biol Macromole 254:127815
11. Barbucci R (ed) (2009) Hydrogels: biological properties and applications. Springer, Milan
12. Matricardi P, Alhaique F, Civiello T (eds) (2016) Polysaccharide hydrogels: characterization and biomedical applications. Pan Stanford Publishing, N.Y.
13. Horkay F, Douglas JF (eds) (2018) Gels and other soft amorphous solids. In: ACS symposium series 1296. American Chemical Society, Oxford University Press, U.K.
14. Marangoni A, Garti N (eds) (2011) Edible oleogels: structure and health implications. AOCS Press, Illinois, USA

15. Patel AR (eds) (2017) Edible oil structuring: concepts, methods and applications. Royal Society of Chemistry, UK
16. Brinker CJ, Scherer GW (2013) Sol-gel glasses: the physics and chemistry of sol-gel processing. Academic Press, N.Y.
17. Wright JD, Sommerdijk NAJM (2018) Sol-gel materials: chemistry and applications. CRC Press, N.Y.
18. Malkin AY, Isayev AI (2022) Rheology: concepts, methods and applications. ChemTec Publishing, Toronto
19. Larson RG (1999) The structure and rheology of complex fluids. Oxford University Press, N.Y.
20. Amjad Z (2007) Water soluble polymers: solution properties and applications. Kluwer Academic Pub., N.Y.
21. Fan L, Yang H, Yang J, Peng M, Hu J (2016) Carbohyd Polym 5(5):309
22. Lin C, Anseth KS (2009) Pharm Res 26(3):631–643
23. Kim S, Kim JH, Kim JO, Ku S, Cho H, Hah P (2014) Macromol Res 22(2):131–138
24. Kannan VN, Narayanan GS (2015) Biomed Res 26:49–51
25. Razzak M, Darwis T, Zainuddin D (2001) Radiat Phys Chem 62:107–113
26. Alexandridis P, Lindman B (2000) Amphiphilic block copolymers: self assembly and applications. Elsevier, N.Y.
27. Hamley I (2005) Block copolymers in solution: fundamental and applications. Wiley. N.Y.
28. He Z, Alexandridis P (2010) Polymer 10(1):32
29. Kabanov AV, Lemieux P, Vinogradov S, Alakhov V (2002) Adv Drug Deliv Rev 54:223–233
30. Alexandridis P, Hatton AT (1994) Colloids Surf A 96:1–46
31. Davedytis M, Zhou JL, Bartlett S (2020) Eng Regener 1:102–113
32. Vasvani S, Kulkarni P, Ratwani D (2020) Int J Biol Macromol 151:1012–1029
33. Harugade A, Sherje AP, Pethe A (2023) React Funct Polym 191:105634
34. Parhi R (2020) Environ Chem Lett 18:577–594
35. Willams PA (2007) Handbook of industrial water soluble polymers. Willams PA (ed). Blackwell, London
36. Phillips GO, Willams PA (2000) Handbook of hydrocolloids. Woodhead Publishing Ltd., Cambridge
37. Lohani A, Singh G, Bhattacharya SS, Hegde RR, Verma A (2014) J Drug Delivery Sci Tech 31:53–64
38. Mokhtari H, Tavakoli S, Safarpour F, Khanazika H, Hamid RBR, Ramakrishna S, Berto F (2021) Polymer 13:1744
39. Zarrintaj P, Manouchehri S, Ahmadi Z, Saeb MR, Urbanska AM, Kaplan DL, Mazafari M (2018) Carbohyd Polym 187:66–84
40. Graham S, Marina PF, Blencowe A (2019) Carbohydrate Polymer 207:143–159
41. Tang Y, Fang T, Lee H-J (2024) Gels 10(6):365
42. Zhou P, Yao J, Cui Y-N, Zhao T, Che J, Yang M, Li Z, Gao C (2022) Gels 8(6):364
43. Shi Y, Ha H, Al-Sudani A, Ellison CJ, Yu G (2016) Adv Mater 28:7921–7928
44. Pu X, Liu M, Sun CJ, Du C, Zhai ZJ, Hu W, Wang ZL (2023) Sci Adv 3:e1700015
45. Kluouda L (2015) Euro J Pharmaceut Biopharmaceut 97:338–349
46. Aoshima S, Oda H, Kobyashi E (1992) J Polymer Sci Part A Polymer Chem 30:2407–2413
47. Katchalsky A, Zwick M (1955) J Polymer Sci 16:221–234
48. Li Y, Young DJ, Loh XJ (2019) Mat Chem Front 3:1489
49. Hamlen RP, Kent CE, Shafer SN (1965) Nature 206:1149–1150
50. Osada Y, Okuzaki H, Hori H (1992) Nature 355:242–244
51. Osada Y, Matsuda A (1995) Nature 376:219
52. Basak S, Gopinath P (2023) Curr Opin Biomed Eng 26:100449
53. Shankar, A., Safronov AP, Mikhnevich EA, Beketov IV, Kurlyandskaya GV (2017) Soft Matter: 3359–3372

Chapter 8
Applications

8.1 Introduction

Various products we use on a daily basis like toothpaste, body wash gels, shampoos, bread, butter, and jam among others are essentially colloidal soft matter materials that contain polymers, oils, water, and surfactants along with other excipients. Surfactants and polymers are inseparable from such colloidal systems and their adsorption on to substrates is an essential process by which it wets a substrate that facilitates processes like adhesion and cleaning. The molecular structure of surfactants and polymers namely its head group charge, hydrophilicity, and hydrophobicity of its backbone and side chains determine its adsorption efficacy [1]. Designing products using a combination of surfactants, polymers, and other excipients for a specific purpose is both an art and a science. The same ingredients used in different amounts and processed through different sequences can yield very different properties that may be used with advantages for diverse applications. Therefore, a deep understanding of the interactions of various components at the molecular level is required while designing a product. In this chapter, various products and their application in fabric wash, personal care, petrochemicals, agrochemicals, pharmaceuticals, food, and lubricants have been briefly discussed.

8.2 Fabric Cleaning

Fabric cleaning is an essential part of any household hygiene that needs to be done often. While hand-washing clothes with a detergent seems to be a prevalent laundry practice in most third-world countries, washing machines have been routinely used in Europe and North America since the early 1950s. Whatever the mode of washing, the efficiency and effectiveness of the washing process depend on the composition of the detergent, the nature of the soil, the characteristics of the fabric, the hardness

A. K. Dutta, *Soft Matter*, https://doi.org/10.1007/978-981-96-0624-5_8

of the water used, the temperature at which washing is done and agitation provided [2–4]. Soil picked up on clothes generally consists of a gooey mixture of organic (oil, sebum, hydrocarbon fragments, dyes) and inorganic (carbon, silica, oxides of iron, clay) materials. These materials have a strong propensity to stick onto fabrics which makes its removal difficult and requires specifically designed products to suit the requirements. In general, formulated detergents consist of surfactants, usually a linear alkylbenzenesulphonate (LABS) or a mixture of LABS and a nonionic usually an ethoxylated lauryl alcohol (C12EO5 or C12EO7), an builder (typically sodium carbonate or a mixture of sodium carbonate and sodium tri-polyphosphate) that lowers or removes the hardness of water through sequestration of Ca^{2+} and Mg^{2+} ions and also imparts an alkaline pH to the washing solution which is essential as LABS does not perform well in neutral and acidic conditions [3–5]. Other ingredients present in the detergent formulation include enzymes, fabric whiteners, color brighteners, and bleaches [3–5]. Perfume and colors although not essential to the washing process are often added to make the product attractive to the consumer. Soil removal from substrates consists of three independent processes (a) roll-up, (b) saponification, and (c) solubilization [3–5] operating simultaneously during the washing process. In the roll-up process, surfactant molecules present in the wash solution get adsorbed at the fabric-water interface and reduce the interfacial tension between oil and water from 30 mN/m to about 10^{-3} mN/m causing the oil drop to roll-up and get detached from the fabric. Figure 8.1a–c shows the removal of oil from a cotton thread wetted with a detergent solution in real-time. Saponification corresponds to sodium ions from sodium carbonate chemically reacts with the oil to form soap that gets dispersed in the wash liquor. Because saponification improves with increasing temperature, cleaning is usually better in warm or hot water. Surfactant molecules present in detergent solution form micelles that solubilize dyes and other organic materials such as oils and aromatics thereby removing stains from the fabric.

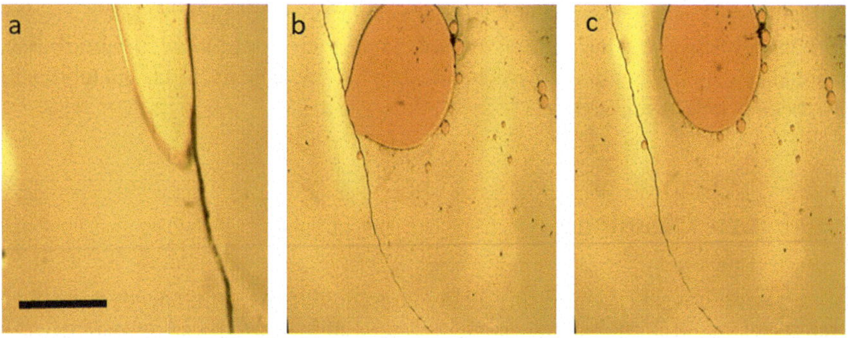

Fig. 8.1 **a**–**c** Shows the roll-up of a mustard oil drop from a cotton thread wetted with a detergent solution in real time as observed through an optical microscope. Images after **a** 120 s, **b** 230 s, and **c** 280 s of adding detergent solution (1 gm/l). Black bar represents 10 μm

8.3 Shampoos

Soft, smooth, beautiful, well-groomed hair is desired by both men and women as it creates an impression of a clean and confident individual. Hair tends to get rapidly soiled due to many reasons namely (a) environmental pollution, (b) moisture and sweat (c) secretion of oily components from the scalp that help dirt particles adhere to it. The basic ingredients of soil are sebum, oil, silica, carbon, oxides of iron, and clay. Because hair consists of a central cortex surrounded by a cuticle that provides shine and a soft feel but is fragile and easily damaged, hair cleaning requires mild surfactants that cleanse well without damaging the protein structure of hair [6–9]. Although linear alkyl benzene sulphonates (LABS) remains the work horse for removing grease and oil from almost any surface, given the high level of irritation it causes to human skin, it is not used in cosmetics and toiletries where mildness and safety are non-negotiable. In general, shampoos contain about 10–20% w/w of sodium or ammonium laurylether sulphates as these are excellent cleansing agents yet mild to human skin. Higher surfactant loading is avoided as it tends to strip the natural layer of oil from the scalp and hair imparting a dry and itchy feeling to the scalp and makes hair dry and rough. Amphoterics, namely betaines, sultaines, and imidazoline derivatives, and alkylpolyglucosides are often used to make milder formulations and deliver consumer-positive cues like foam [1, 2, 6–9]. Although soaps were used for hair in the past, today it is hardly ever used as its use tends to swell the hair shaft owing to their high pH making hair harsh and unmanageable. Hard water tends to deposit calcium and magnesium soaps on hair leaving behind a rough, dry, and scruffy appearance.

8.4 Body Wash Gel

Although toilet bath soaps have been around and for more than a century and continue to be so, recent changes in that trend have been observed as a large fraction of the users today prefer body wash gels. These gels provide the convenience of use, superior feel, lesser amount of water to rinse off the foam compared to soap not to mention the appearance of bars getting soggy on being left wet or cracked on drying. Body wash gels usually comprise a mild surfactant namely sodium laureth sulphate (SLES), glycerin, cocoamidopropyl betain, PEG-40 hydrogenated Castor oil, PEG-7 glyceryl cocoate, benzyl salicylate, essential oils (lemon grass oil, geraniol) and perfume. As a rheology modifier hydroxypropyl cellulose and sodium chloride are used to thicken the product to make a flowable gel such that it is readily pourable and pumpable [6, 7, 10–12].

8.5 Enhanced Oil Recovery from Oil Wells

Limestone and sandstone are sedimentary rocks that hold large volumes of oil trapped in their micro capillary network.

The oil recovery process from these reserves usually consists of three phases. The first step called the primary process involves direct drilling into these rocks that allows part of the oil to flow out on its own due to pressure and the yield depending on the oil field is about 20% of the total oil entrapped in these rocks. This low yield is due to several factors but mainly due to (a) high viscosity of the oil in the micro-capillaries that make their flow slow and recovery difficult (b) adhesion between the oil and the walls of the pores, (c) tortuosity of the microporous network and (d) width of the micro-capillaries. In the second phase or secondary process, steam and gases (carbon dioxide or nitrogen) are pumped into the well which helps recover another 10–20% of the trapped oil. Elevation of temperature causes a lowering of the oil viscosity that encourages its flow under pressure. The third step or tertiary process also called the enhanced oil recovery (EOR) step involves the use of chemicals (surfactants, polymers, or their mixtures) for further recovery of the oil from the oil field. Usually, this network of pores in the rock contains brine, oil, and gases. Prolonged contact between the oil and the substrate of the pore modifies its wettability that results in strong adhesion between the oil and the pore walls that makes recovery difficult. Addition of surfactants lowers the interfacial tension between oil and water from about 30 to 10^{-3} mN/m that loosens up the slug as shown in schematic Fig. 8.2 making extraction possible and easier [13, 14].

8.6 Metal Working Fluids

Intense heat is generated during metal cutting operations due to friction between the cutting blade and the metal. Sometimes the heat generated may be so high that the metal itself melts which is undesirable. Therefore, a cooling system capable of rapidly absorbing the heat generated and simultaneously providing lubrication between the cutting tool and the metal block is required [15]. While water owing to its high specific heat capacity can provide cooling, it does not provide lubrication. On the other hand, oils are known to be good lubricants but the use of oil directly had limited success as cooling was ineffective. Hence use of both water and oil simultaneously through formulations that provide both cooling and lubrication is desired. These formulations include emulsifiers, lubricants, dispersants, wetting agents, and corrosion inhibitors. The surfactants are so chosen that they must get strongly adsorbed at the metal surface and form a hydrophobic layer that will provide lubrication between the cutting or grinding surface and the metal. Long chain surfactants namely C_{12}–C_{18} phosphate esters are used because they interact with the metal surface forming tightly packed monolayers that provide the necessary lubrication. Other surfactants preferred are ethoxylated derivatives of sulphates, sulphonates, and phosphates as these enhance

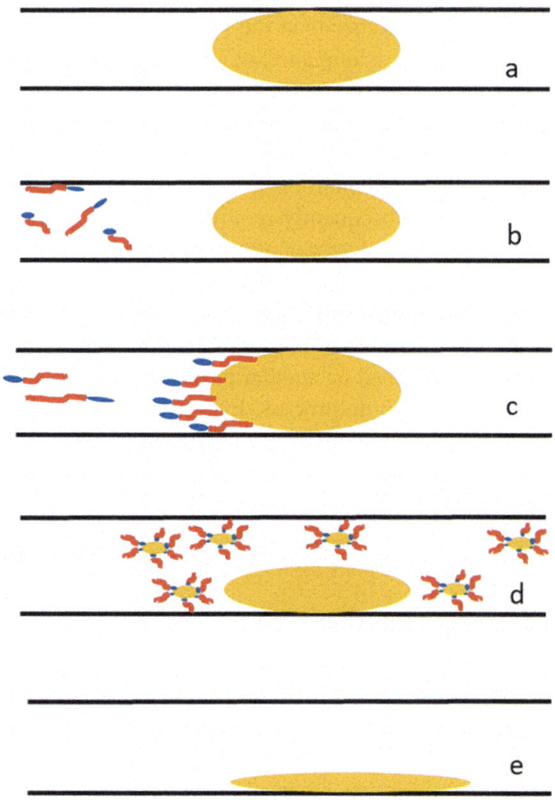

Fig. 8.2 Schematic of a **a** hydrocarbon oil in the capillary in a sedimentary rock, **b** the capillary is flooded with a surfactant solution, **c** adsorption of surfactant on oil drop, **d** partial emulsification of the oil in capillary, **e** residual oil sticking to walls of capillary

emulsification and adsorption of the surfactant on the substrate. Branched nonionics with C_8–C_9 alkylated phenols and or branched C_{10}–C_{13} nonionics with EO_n where $n = 5$–10 are often used. With friction, temperature rises, and above the cloud point the emulsion switches over from oil in water (O/W) to water in oil (W/O) type emulsion, releasing oil onto the metal surface that ensures good lubrication.

8.7 Paper Processing

In the paper manufacturing industry, newsprint is often recycled. Removal of ink from wet pulp slurry is essential before the paper can be fully reprocessed. It is for this purpose surfactants are added to the wet slurry so that the carbon particles may be dispersed and removed from the slurry. Branched alkylchain nonionics are preferred over other surfactants as they can provide better wetting and reduction of surface tension and interfacial tension. Redeposition of particles is prevented by adding high molecular weight (M.W. 5000–10,000) EO/PO/EO block copolymers with 20–80% EO substitution [16]. The nonionics are so chosen that their cloud temperatures are about 5–10 °C higher than the pulp temperatures that ensures good dispersion. In USA, both the washing and flotation processes are used in practice.

In the flotation process, a small amount of a long chain soap is added that reacts with calcium and magnesium ions to form salts that are adsorbed on the carbon particles such that the long chain remains protruded in water and gets easily attached to the hydrophobic microenvironment provided by the bubbles helping them to keep afloat and be subsequently removed. Because fatty acid salts tend to adsorb at the fiber surface, an inferior quality paper is generally produced which is not intended. Hence surfactants of the type C_{16}–O–$(EO)_{10-12}$–$(PO)_{4-6}$H are used that adsorb better on polar ink particles with the hydrophobic chains pointing toward water making the ink particles hydrophobic [16]. When the temperature is lowered, the longer PEO chains get adsorbed on smaller particles acting as steric stabilizers and are removed through the washing process. End-capped PO surfactants with low viscosities and random chain compounds are mostly liquids and therefore preferred. End-capped PO alcohols are in general superior to the EO alcohols and hence used extensively [16].

8.8 Agrochemicals

The use of pesticides is an essential and unavoidable part of crop protection and pest management activity across the globe. Various pesticide actives are necessary to contain different pest attacks during different seasons of the year even for a single crop. Several formats are available namely emulsifiable concentrates (EC), suspension concentrates (SC), and oil dispersions (OD) while common solid formats used are dry flowable (DF) and water dispersible granules (WDG) [17]. Briefly, ECs consist of a mixture of active, surfactants and an organic solvent. When mixed with water, an oil-in-water (O/W) type emulsion is formed with the active localized in the oil droplet. On spraying, these droplets get deposited on the foliage and over time water evaporates leaving the actives on the leaves. SC's consist of a mixture of finely divided solid active, surfactants and/or polymers in water that makes an emulsion that is sprayed on the foliage. ODs are diluted with water or oil and sprayed on the crop that leaves behind a very fine layer of the active that protects the foliage from pest or fungal attack. Considerable research has shown that effectiveness of a pesticide spray not only depends on the right choice of active for a particular crop and pest but on other factors like size of droplets, surface coverage of the spray target, retention of the spray droplet, time duration of the droplet residing on the foliage. Considering a pesticide-laden drop on a leaf to be a spherical cap sitting on a substrate it has been calculated by Ford [18] and Furmidge [19]. The ratio of the surface area covered to the diameter of the drop also known as the "spreading factor" is given by

$$SF = \left[\frac{4\sin^2\theta}{(1-\cos\theta)^2(2+\cos\theta)}\right]$$

where θ is the contact angle made by the pesticide solution on the leaf substrate. It is evident from the plot of SF versus θ that the spreading factor increases rapidly with decreasing θ and is a maximum when $\theta < 35°$.

Similarly, the amount retained on the substrates has been calculated to be.

$$R = \theta_M \left[\frac{\gamma_{la}(\cos\theta_R - \cos\theta_a}{\rho} \right]^{\frac{1}{2}}$$

where θ_a and θ_R corresponds to the advancing and receding angles of the droplet on the leaf substrate and $\theta_M = \frac{\theta_a + \theta_R}{2}$. These relations are independent of the orientation of the substrate which in a real-life situation should correspond to a distribution function for a tree which would again depend on the geometry of that tree. Because of the retention (R) and spreading factor (SF) dependence on γ_{la} and θ, it is implicit that formulations have to be compared and optimized to deliver maximum benefits. However, contact angles are known to depend on the surface texture of the substrate and hence would depend on the leaf surface. Since many approximations have been made to derive the above result, various local factors also play a crucial role.

8.9 Foods

Taste is perhaps the single most important criteria that makes a food product succeed or fail commercially. Tailoring physico-chemical properties of food materials namely the droplet size, type of fat crystals, their densities, and their rheological properties that alter textures and hence taste and mouth feel of foods is crucial. Margarines are typically water in oil emulsions [20, 21] where the solid content mainly fats and oils varies between 5 and 25% by weight. The water droplet size distribution is known to be very broad. While small droplets of diameter 2–4 μm provide protection against mold growth, larger droplets with diameter typically 10–20 μm provides better flavor through release of flavor components in the mouth [22]. Fat crystals exist in two forms β and β'. While the β variety has a higher melting point and forms large plate shaped crystals that are about 20 μm in length, the β' crystals are small needles that are about 1 μm long that provides a fine feel of the product, large β type crystals produces a grainy texture. The relative abundance of the crystals determines the final quality and stability of the product [22–24]. Addition of partially hydrogenated sunflower oil or low euricic rape seed oil increases the abundance of β' crystals hence the graininess of the product. Under pressure, a transition from the β' to the β form of crystal reportedly leads to a separation of oil and the semi-solid phase which is unwanted by the consumer. Adding 0.3% w/w sorbitan tri-stearate prevents such transformations and such studies are essential in the food industry to maintain and manage quality and consumer aspirations.

8.10 Pharmaceuticals

Many anticancer, antitumor, and psychotropic drugs used in pharmaceutical formulations are actually unstable and insoluble in water. These drugs are delivered using different vehicles namely as microemulsions and microencapsulates. Recently, delivery systems consisting of nanoparticles and vesicles have also been used with great success. In this section, we will briefly discuss several drug delivery systems.

Microencapsulates: Glizipide and Indomethalin have been incorporated in alginate/sodium cmc/HPMC microspheres that are delivered orally [25]. Starch microspheres have been used to deliver insulin and/or gentamycin through the nasal route in the form of aerosols [26].

Microemulsions: Cyclosporin A, Itraconazole, and Celecoxib have been formulated as microemulsions for oral intake constitute a self-emulsification drug delivery system that is more effective and efficient in contrast to drugs delivered through the oral solid or lipid route [27–30].

Nanoemulsions: Antifungal drugs fluconazole and antibiotic chlorhexidene have been delivered using a PLA-PEG/CCT and a PCL/Labrafac hydrophilic WL-1219 polymer systems, respectively [31, 32].

Micelles: Doxorubicin is an anticancer drug that is administered intravenously can accumulate in tissues that may lead to high toxicity in these tissues that needs to be avoided. This is achieved by using doxorubicin with block copolymer micelles. Because micelles have a non-polar core, the sparingly soluble drug is removed from the blood stream and transferred to the micelle thereby reducing its toxicity ensuring a longer and sustained delivery of the drug. Interestingly, studies by Hofheinz, Gradisher and Mrozek [33–35] has demonstrated that doxorubicin delivered through a liposomal system is better tolerated by breast cancer patients.

8.11 Superhydrophobic Textiles

Non-wettable and non-stainable fabrics have become popular in recent times as these do not tend to get dirty and preserve their "new look" for long periods of time. Chemical vapor deposition of perfluoroalkylmethacrylate on fabrics increases the contact angle from 0° (complete wetting) to about 175° and a sliding angle of 25° for a 20 mg water droplet. Wool, polyester, and cotton treated with silica sol of cohydrolyzed tetraethyl orthosilicate (TEOS) and fluoroalkylsilane with ammonium hydroxide yield a contact angle of 170° and siding angle of less than 7° [36]. Xu et al. [37] has demonstrated that coating fabrics with silica and zinc oxide nanorod arrays made cotton hydrophobic with contact angle greater than 150°. Zhu [38] demonstrated that cotton fabrics treated with a silica sol prepared through hydrolysis condensation of tetraethoxysilane under alkaline conditions and hexadecyltrimethoxysilane produce a water and stain-resistant finish that remains functional even after 30 laundry wash cycles.

8.12 Synthesis of Inorganic Nanoparticles

Since the late 1980s, there has been considerable interest in the synthesis, characterization, and application of semiconductor and metallic nanoparticles. Interest in these systems arises from the size-dependent optical and electronic properties of these nanoparticles that find potential applications from high-end optoelectronic and photonic devices to cancer therapy and bio-imaging [37–41]. Although various methods of generating nanoparticles have been reported [42–49] perhaps the simplest and most elegant of them all involves producing the particles in reverse micelles and in microemulsions developed by Pileni et al. [43–46]. It was demonstrated that nanoparticles may be synthesized by allowing a reaction between two reverse micellar solutions both having an aqueous core containing metallic (M^+) the other an anionic (A^-) group (hydroxyl, sulphate, sulphide, etc.).

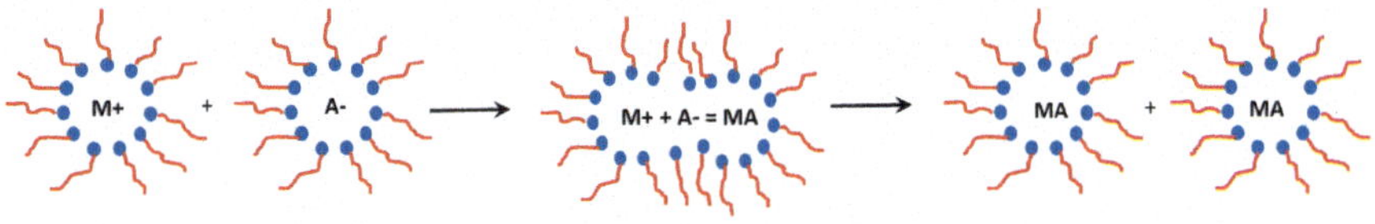

The reverse micelles consisted of iso-octane as the bulk solvent while water was the core and the surfactant used was AOT (2-ethylhexylsulfosuccinate). Detailed studies [43–49] have confirmed that the single parameter determining the size of the nanoparticles depended on w, where w corresponds to the ratio of the molar concentration of water [H_2O] and the molar concentration of 2-ethylhexylsulfosuccinate [AOT]. This technique is unique in that it allows control over the crystal size in contrast to that made from mixing of solutions in a container as there is no constraint to limit the La Mer [50] nucleation and growth process and hence large crystals are produced. Nanoparticles of iron oxides prepared by the above method have been used in medical diagnostics namely MRI.

8.13 Polymeric Hydrogels in Ultrasonography

Using acoustic waves to detect abnormal growth in tissues specifically tumors is a safe and non-invasive means of diagnosis that is used extensively by the medical fraternity. Here an ultrasound device with a scanning head that contains both an ultrasound emitting probe and a detector is brought in contact with the section of the body that needs to be examined. The ultrasound probe emits the signal that penetrates the tissues and the attenuated reflected signal is detected by the detector that is used to digitally construct the image from which the ultrasound has bounced. A key difficulty of the entire process is that air space between the scanner head and the skin causes a large fraction of the ultrasound waves to get reflected which interferes with the image. To prevent this a thin layer of hydrogel is used to fill the air gaps between the scanning head and the skin surface. The usual hydrogels used are either a polyacrylate or a polyacrylamide gel containing about 90% water and a very small amount (< 0.5%w/w) of an antimicrobial preservative like 1,2 dibromo-2,4 dicyanobutane. These gels help make excellent contact between body surface and the scanning head which improves the signal quality and image contrast [51].

References

1. Rosen MJ (1989) Surfactants and interfacial phenomena, 2nd edn. Wiley-Interscience, N.Y.
2. Farn RJ (eds) (2006) Applied theory of surfactants, chemistry and technology of surfactants. Blackwell Publications, London
3. Broze G (eds) (1999) Handbook of detergents, part A, properties. Marcel-Dekker, N.Y.
4. Lange KR (1994) Detergents and cleaners. Hanser, Munich
5. Jakobi G, Lohr A (1986) Detergents and textile washing. VCH, Weinheim
6. Barel AO, Paye M, Maibach HI (2001) Handbook of cosmetic science and technology. Marcel-Dekker, N.Y.
7. Rosen MR (eds) (2005) Delivery system handbook for personal care and cosmetic products, technology, application and formulations. William-Andrew Inc., N.Y.
8. Rhein LD, Schlossman M, O'Lenick A, Somasundaran P (eds) (2007) Surfactants in personal care products and decorative cosmetics. CRC Press, Florida
9. Draelos ZD (2005) Hair care: an illustrated dermatologic handbook. Taylor-Francis, U.K.
10. Tadros TF (2018) Formulation science and technology, vol 4. DeGruyter, Berlin
11. Tadros TF (2017) Formulations in cosmetic and personal care. Degruyter, Berlin
12. Tai LHT (2000) Formulating detergents and personal care products: a guide to product development. AOCS Press, USA
13. Sheng JJ (2011) Modern chemical enhanced oil recovery: theory and practice. Elsevier, USA
14. Dai C, You Q, Zhao M, Zhao G, Zhao F (2023) Principles of enhanced oil recovery. Springer, Singapore
15. Astakhov VP, Joksh S (2012) Woodhead Publishing Limited. Oxford
16. Ek M, Gellerstedt G, Henriksson G (2009) Paper chemistry and technology, vol 4. De Gruyter, Berlin
17. Tadros TF (2009) Colloids in agrochemicals: colloids and interface science, vol 5. Wiley-VCH, Berlin
18. Ford RE, Furmidge CGL (1967) J Sci Food Agrochem 18:419–428
19. Furmidge CGL (1962) J Sci Food Agrochem 13:127–140
20. O'Brien RD (2009) Fats and oils: formulating and processing for applications. CRC Press, Florida
21. Jaynes EN (1985) Applications in the food industry in encyclopedia of emulsion technology, applications, vol 2. In: Becher P (ed). Marcell-Dekker, N.Y.
22. Friberg SE, Larsson K, Sjoblom J (eds) (2005) Food emulsions. Taylor & Francis, USA
23. Mathiowitz E, Jacob JS, Jong YS, Carino GP, Chickering DE, Chaturvedi P, Santos CA, Morella C, Basset M, Vijayraghavan K (1997) Nature 386:410–414
24. Farraj NF, Johansen BR, Davis SS, Illum LJ (1990) Controlled Release 13:253–261

25. Mueller EA, Kovarick JM, van Bree JB, Tetzloff W, Grevel J, Kutz K (1994) Pharm Res 11:301–304
26. Woo JS, Song YK, Hong JY, Lim SJ, Kim CK (2008) Eur J Phar Sci 33:159–165
27. Garti N, Avrahmi M, Aserin A (2006) J Colloid Interface Sci 299:352–365
28. Zhang P, Liu Y, Feng N, Xu J (2008) Int J Pharm 355:269–276
29. deAsis DN, Mosqueira VCF, Vilela C, Andrade MS, Cardoso VN (2008) Int J Pharma 349:152–160
30. Nhung DT, Freydiere AM, Constant H, Falson F, Pirot F (2007) Int J Pharm 334:166–172
31. Hofheinz RD, Gnad-Vogt SU, Beyer U, Hochclaus A (2005) Anticancer Drugs 16:691–707
32. Gradishar WJ (2005) Breast Cancer Res Treat 89:S17–S26
33. Mrozek E, Rjoases CA, Allen J, Hade EM, Shapiro CL (2005) Ann Oncol 16:1087–1093
34. Wang H, Fang J, Cheng T, Ding J, Qu L, Dai L, Wang X, Lin T (2008) Chem Commun: 877–879
35. Xu B, Cai Z, Wang W, Ge F (2010) Surface Coating Tech 204:1556–1561
36. Zhu Q, Gao Q, Guo Y, Yang CQ, Shen L (2011) Ind Engg Chem Res 50:5881–5888
37. Fahrner WR (ed) (2004) Nanotechnology and nanoelectronics, materials, devices and measurement techniques. Springer, N.Y.
38. Mitin VV, Kochelap VA, Stroscio MA (2009) Introduction to nanoelectronics. Cambridge University Press, Cambridge
39. Hosokawa N, Nogi K, Naito M, Yokoyama T (2012) Nanoparticle technology handbook. Elsevier, N.Y.
40. Gao X, Cui Y, Levenson RM, Chung WLK, Nie S (2004) Nature Biotech 22:969–976
41. Weissledee R (2006) Science 312:1168–1171
42. Liveri VT (ed) (2006) Controlled synthesis of nanoparticles in microheterogeneous systems. Springer, N.Y.
43. Pileni MP (2011) J Phys Condensed Matter 23:503102
44. Pileni MP (1989) Structure and reactivity in reverse micelles. Elsevier, Amsterdam
45. Pileni MP (1997) Handbook of surface and colloid chemistry. In: Birdi KS (ed). CRC Press, N.Y.
46. Pileni MP (1993) J Phys Chem 97:6961–6973
47. Petit C, Lixon P, Pileni MP (1990) J Phys Chem 94:1598–1603
48. Qi L (2006) Encyclopedia of surface and colloid science, vol 2, 6183–6207. In: Somasundaran P (eds) CRC Press, N.Y.
49. Cizeron J, Pileni MP (1995) J Phys Chem 99:17410–17416
50. La-Mer VK (1952) Ind Eng Chem 44:1270–1277
51. Poltawski L, Watson T (2007) Ultrasound Med Biol 33:120–128

The manufacturer's authorised representative in the EU is Springer Nature Customer Service Centre GmbH, Europaplatz 3, 69115 Heidelberg, Germany. If you have any concerns regarding our products, please contact ProductSafety@springernature.com

Printed and bound by CPI Group (UK) Ltd, Croydon, CR0 4YY
01/10/2026
02225945-0007